Computer Simulation of Power Systems

Programming Strategies and Practical Examples

by D. James Benton

Forward

There are many texts and examples devoted to computer simulation of a single process or component. This text focuses on entire systems. From the earliest days of my career, I have developed computer simulations of various sorts. Many of these have been time sequence based; that is, modeling the response of a system to changes over time. Most often, the source of change has been the environment or weather-related. In this book we explore how to build models, acquire, structure, and extend weather data, perform, and analyze the results of complex simulations. We consider rivers and lakes, hydroelectric, conventional and combined cycle power generation, nuclear, and solar plants. All of the examples are based on actual systems and all of the software is available free online.

All of the examples contained in this book,
(as well as a lot of free programs) are available at...
https://www.dudleybenton.altervista.org/software/index.html

i

Table of Contents   page

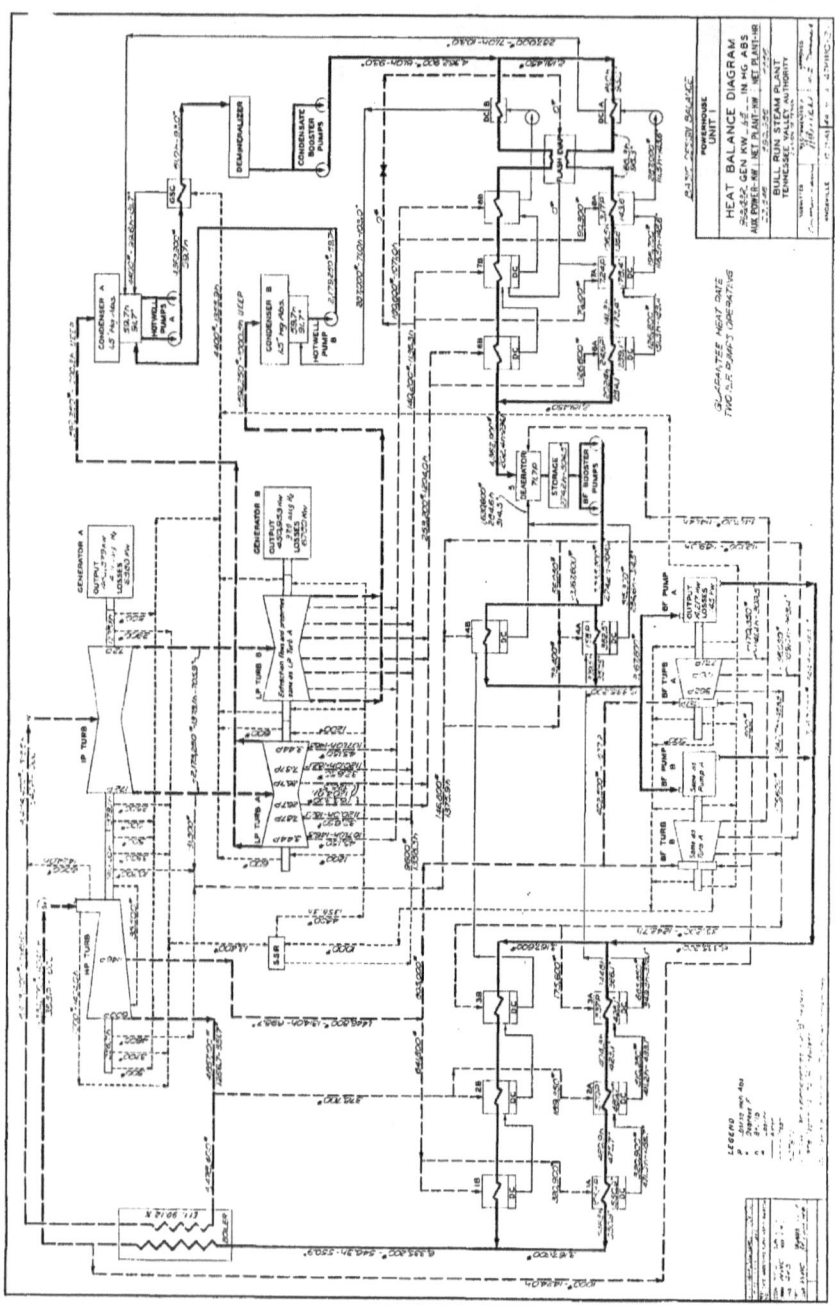

Chapter 1. Browns Ferry Nuclear Plant

We begin with a complete working system model, see how it runs, and then consider the components that are combined to create the whole simulation. I was tasked with developing this system model in 1982, when FORTRAN was the dominant programming language available, although I had worked extensively with assembler before this time. The Tennessee Valley Authority (TVA) had built a nuclear power plant in Northern Alabama adjacent to Wheeler Reservoir on the Tennessee River (see Chapter 2 for more details). The plant consisted of three General Electric boiling water reactors (BWRs).

The central building is the powerhouse. The tall, slender stack is the vent shaft (unique to BWRs). The cooling water channel is in the foreground along with one of the cooling towers. When design began on this plant, it was assumed that the heat rejection (i.e., condenser cooling water) would be discharged to the river through three multi-port diffusers.

Each diffuser pipe contains a 183 m (600 foot) long discharge section with ports spaced in alternating columns of six and seven 5.1 cm (2 inch) diameter holes, situated 15 cm (6 inches) apart on center, both vertically and horizontally, totaling about 7,000 ports per diffuser. The ports face downstream, and depending on the location in the port column, contain a discharge angle between

1

24°and 45°from horizontal. The discharge sections of the diffusers are situated in succession across the 550 m (1800 foot) width of the main channel.

The upstream diffuser, for Unit 2, is 6.25 m (20.5 feet) in diameter and is situated with the discharge section in the south side of the main channel. The middle diffuser, for Unit 1, is 5.79 m (19.0 feet) in diameter and is situated with the discharge section in the middle of the main channel. The downstream diffuser, for Unit 3, is 5.18 m (17.0 feet) in diameter and is situated with the discharge section in the north side of the channel. Conceptually, the diffusers look something like:

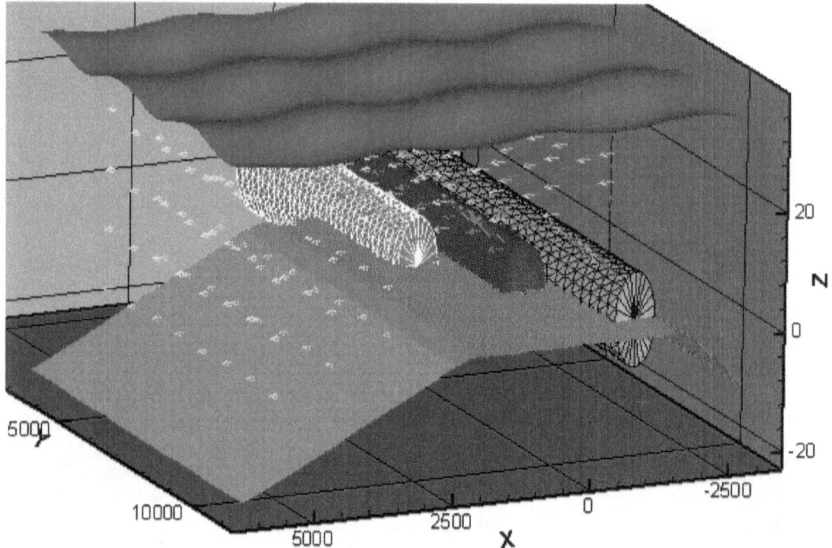

Before the plant became operational, the state of Alabama imposed thermal water quality standards (maximum downstream mixed temperature and plant-induced temperature rise) plus the U.S. Nuclear Regulatory Commission (USNRC) imposed a maximum cooling water temperature. These restrictions necessitated the installation of six large mechanical draft cooling towers, pictured previously.

It was originally thought that the cooling towers would only be needed occasionally, however, this was not the case. I developed this computer model (bfsim.c), which can be found in the online archive in folder examples\bfsim, to quantify how often the cooling towers would be needed and what the impact on plant performance might be. This study revealed several things, including: weather and river conditions would significantly impact plant operation. We will get to the model details shortly, but first we will consider some of the results in graphical form for plant operation without cooling towers.

The following figure shows the wet-bulb temperature (which directly impacts cooling tower performance and shown in light cyan), cooling channel discharge temperature (magenta), upstream (blue) and downstream (red) river temperature and plant-induced river temperature rise (dark red). The limits are shown as dotted lines of matching color.

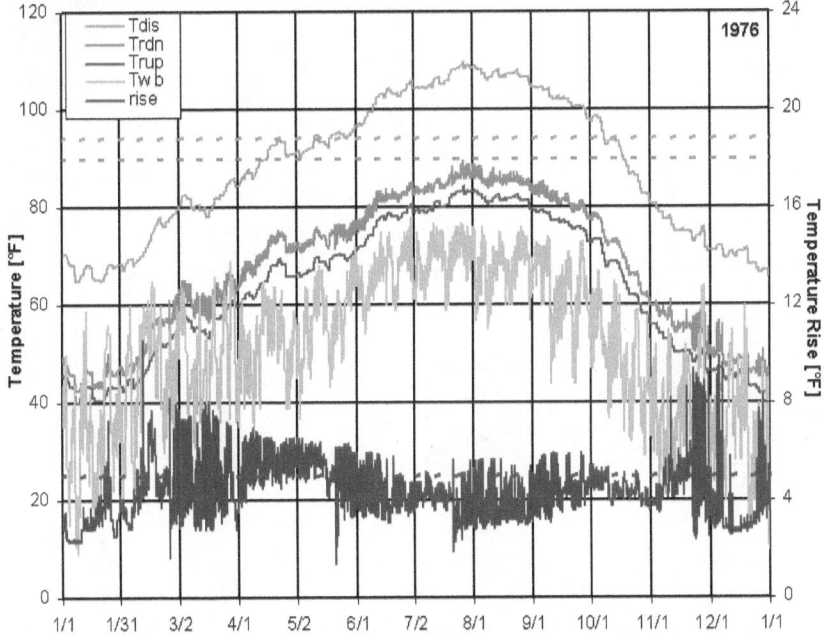

The wavy magenta curve is above the dotted magenta line from mid May through mid October. This is a definite problem, as this represents a safety limit set by the USNRC. The wavy red line doesn't rise above the dotted red line, but that's only true in cool years, which this first available one happened to be. The dark red (plant-induced river temperature rise) curve exceeds the dotted line for much of the year. The conclusion is clear: even in this rather mild year (1976), the cooling towers would be necessary.

The plant-induced temperature rise depends most heavily on the river flow rate, which is lowest in March, late November, and early December. This flow pattern arises from flood control objectives combined with seasonal rainfall and was implemented before the plant became operational. One of the conclusions of this study was that future operations would need to consider the plant, as the previous operations frequently resulted in zero and even negative flow past the plant due to *sloshing*.

Note that the upstream river temperature (blue) curve is mostly above the wet-bulb (light cyan) curve. This pattern arises from the fact that the river

3

absorbs and retains solar heat much more readily than the atmosphere. The response time (wiggliness) of the atmosphere (i.e., wet-bulb) is much greater than that of the river, due to the thermal mass. The hottest river temperatures in this early data set were in 1980.

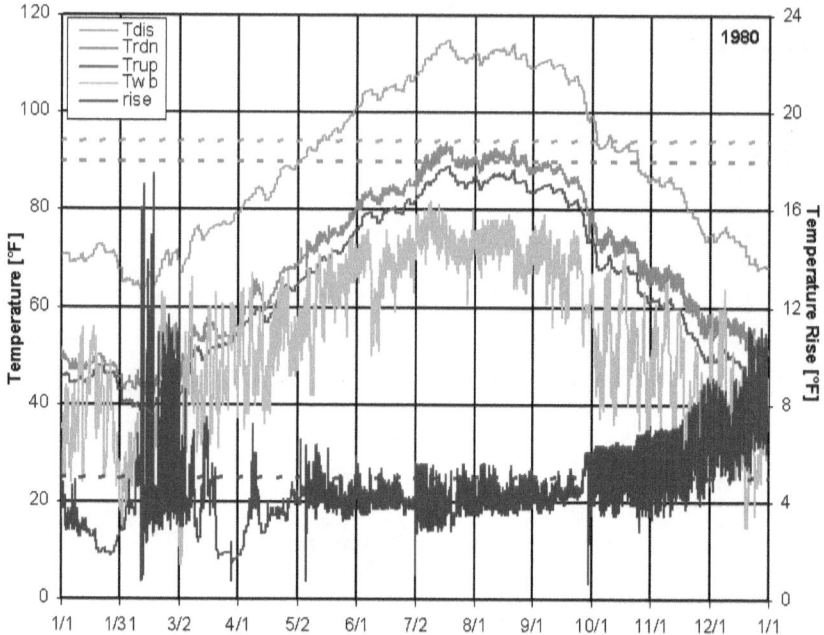

In this figure we see that the downstream river temperature (red curve) is at or above the dotted red line (water quality limit imposed by the state of Alabama) from mid July through late August. At least for this particular year, the river flows during this time were higher than average, as reflected in the smaller variations in the plant-induced temperature rise (dark red) curve. The flows in 1980 were right in the middle between highest and lowest for the ten-year period of 1976-1985.

The lowest flows during this time occurred in 1981. Various studies have continued for this same plant, at least through 2015. Collection of data has also continued and, as it has turned out, 1980 was not the hottest year, nor was 1981 the driest. Still, installation and operation of the cooling towers made a significant impact on the waste heat rejected to the river, as shown in the next figure; however, even this measure was not enough to meet the environmental regulations set forth by the state of Alabama or the thermal safety limit imposed by the USNRC.

The next figure shows these same calculations for 1981 without cooling tower operation and the figure after that shows a comparison of plant-induced

4

river temperature rise with and without cooling towers for the 1981 flows and meteorology.

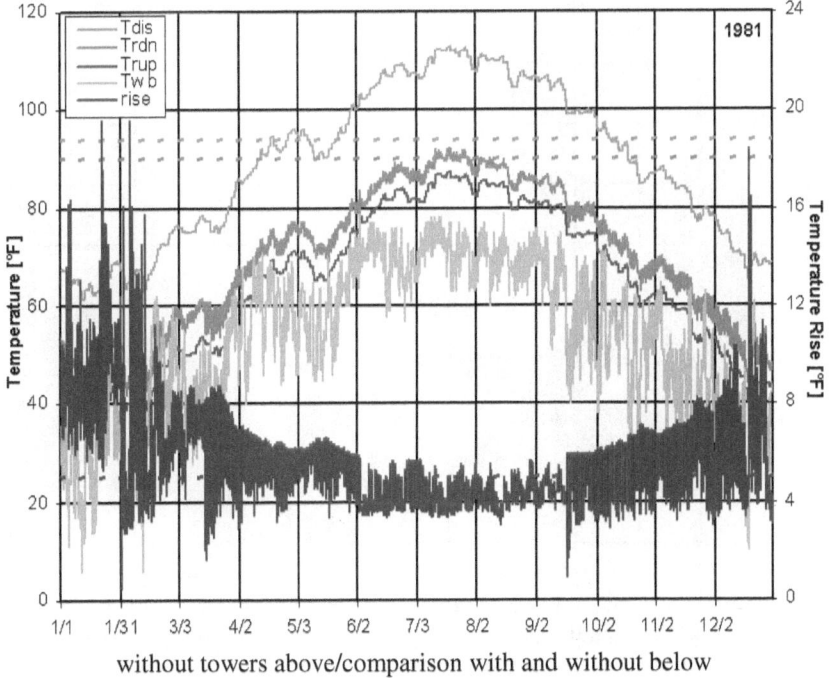

without towers above/comparison with and without below

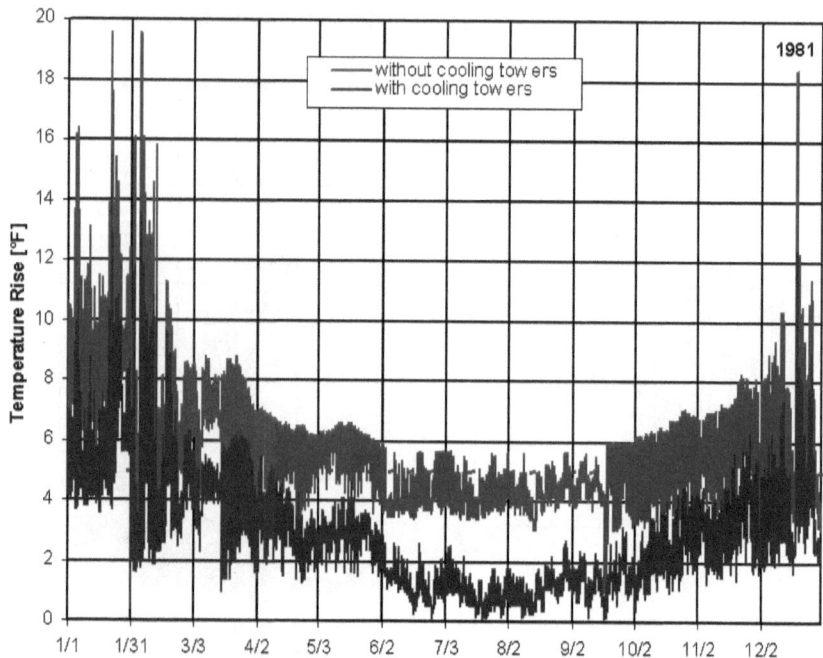

The dark blue line in the preceding figure is often above the dotted dark red one ($\Delta T=5°F$), especially in January and February as well as mid to late December. This indicates that the water quality standards could not be met, even with cooling towers, if the plant were operated at full power. This conclusion illustrates the importance of such studies. Clearly, something additional must be done to utilize and recoup the investment in this multi-billion dollar asset and still protect the environment.

Browns Ferry Nuclear Plant was further complicated by a series of accidents. There was an electrical fire in 1975 that damaged Unit 1. The cooling towers were made of redwood. One of the main cooling towers burned down on May 10, 1986. A smaller cooling tower burned down on May 24, 1996. As even the original cooling towers in new condition were not capable of dissipating all of the waste heat generated by the three units operating at full load, disagreements continue to this day over the best approach to address the situation. Computer modeling (i.e., simulation based on historical data) remains an essential part of this planning process.

<u>System Components</u>

Now that we have seen what the model can be used for, let us consider the parts (i.e., components) that make up the whole. The calculations include: river flows, thermal cycle heat rejection, condenser performance (that feeds back into the heat rejection), cooling tower range, multi-port diffuser dilution, and the principal cooling water flows. We will consider each of these and how they combine to arrive at system response.

This section of the Tennessee River is bounded above by Guntersville Dam and below by Wheeler Dam, which combined form Wheeler Reservoir. Flow past the plant is not measured, but flow through each of the dams is. The dam releases and/or hydroelectric generations are logged, archived, and forecasted. Flow along the reservoir and at the plant is calculated using a transient routing (open channel) model. The MacCormack calculation scheme is used.[1] This method is presented in the reference below and compared to a more common one by Ferrick, a colleague, and Waldrop, my supervisor at the time this work was conducted and also the Assistant Director of the Laboratory.

The heat rejection calculation is based on the backpressure correction curves and heat balances supplied by the manufacturer (General Electric). The curves show the change in heat rate (thermal heat input divided by net electrical power output) with condenser pressure and power level (25%, 50%, 75%, and 100%). A bivariate curve fit is used, based on the saturation pressure of steam. The curves are shown in this next figure.

---

[1] Ferrick, M. G., and Waldrop, W. R., "Two Techniques for Flow Routing with Application to Wheeler Reservoir," TVA Report No. 3-519, 1977. http://dudleybenton.altervista.org/pub/TVA3-519.pdf

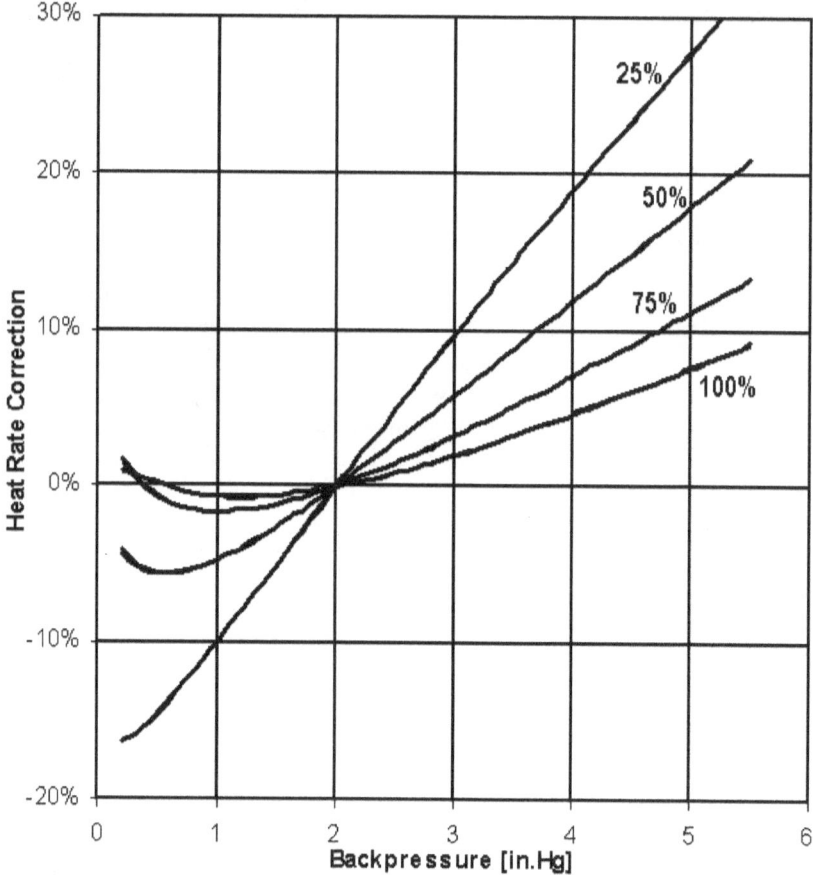

The condenser performance is calculate using the Heat Exchange Institute standards, which includes a water temperature correction factor to account for changes in thermal conductivity and viscosity. The latest version is the *Heat Exchange Institute Standards for Steam Surface Condensers*, 11th Ed. (2012). As the condenser pressure depends on how much heat is rejected and the heat rejection depends on the condenser pressure, this is an iterative calculation:

```
for(iter=0;iter<5;iter++)
  {
  dtc=fhrej/3600./62.4/qcond;
  tsat=tin+dtc/(1.-exp(-ntu));
  tbp=psat(tsat)/0.491115;
  epcf=((1.1823-0.312/duty)*(tbp-2.)+(2.7348/duty-
  1.543))*(tbp-2.);
  epcf=fmax(-50.,fmin(100.,epcf));
  fhrej=(1.+epcf/100.)*fhrej0;
  }
```

Condenser performance is most often provided by the manufacturer as a family of curves. The following arise from the HEI standards:

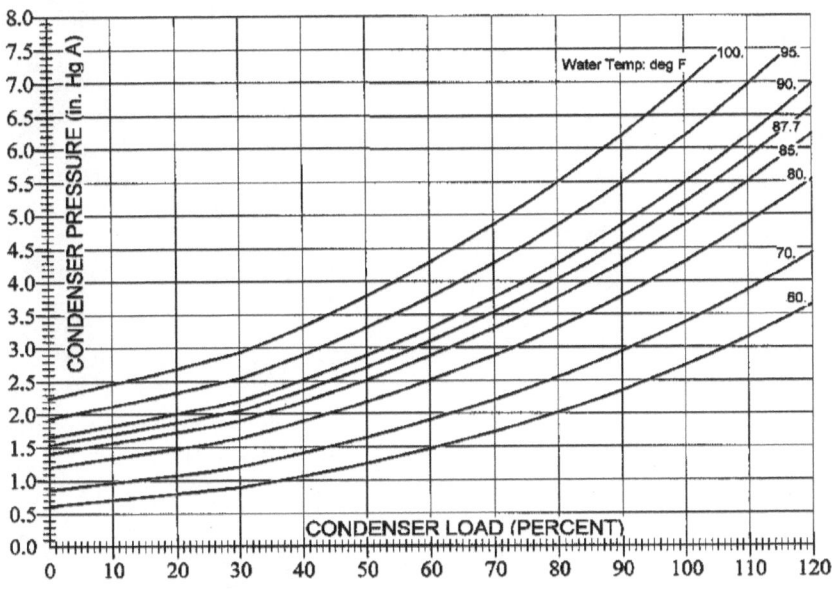

condenser curves above/cooling tower curves below

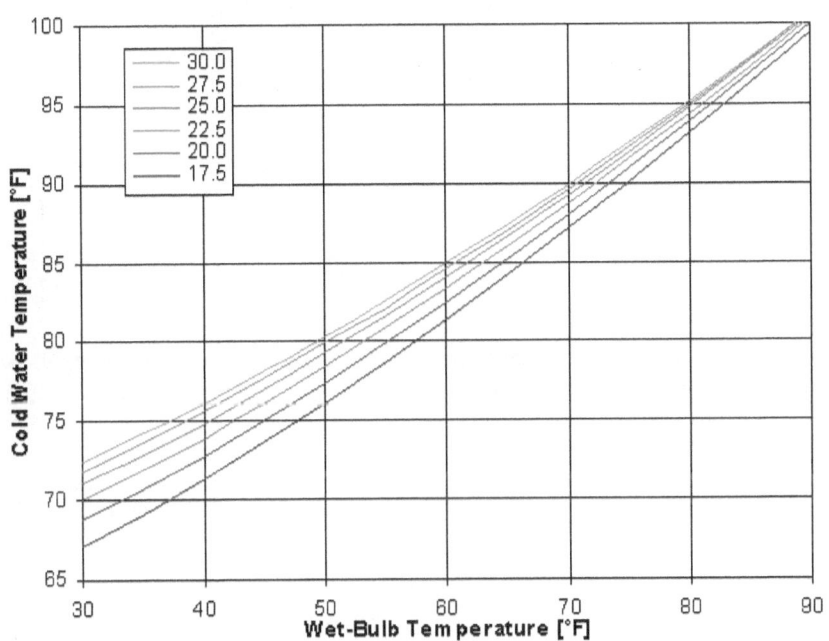

Cooling tower performance curves are most often drawn as cold water temperature (i.e., exiting the tower) vs. ambient wet-bulb temperature for various values of cooling range (i.e., hot entering water temperature minus cold exiting water temperature). The cooling tower manufacturer, Ecodyne, provided these curves.

The multi-port diffuser is both unusual and also somewhat unique to this design. The TVA Engineering Laboratory was a pioneer investigating these devices, along with the Iowa Institute of Hydraulic Research. This particular design is briefly described by Stolzenbach.[2] A more detailed analysis of these devices was provided by Jirka and Harleman.[3]

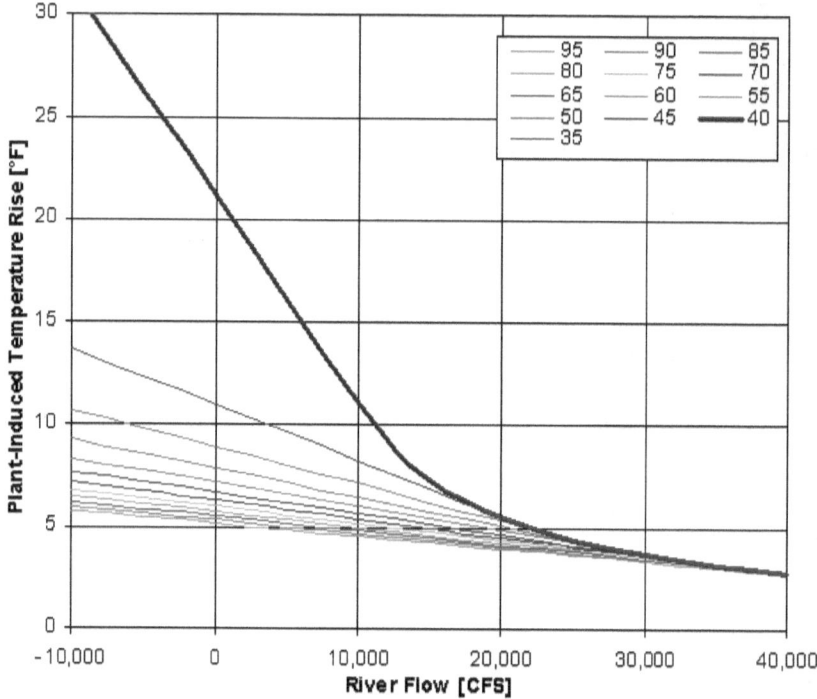

Diffuser performance depends on several variables so that a simple graph is not adequate to represent the range of conditions. The preceding figure is based

---

[2] Stolzenbach, K. D., "Estimation of Water Temperature Increases In Wheeler Reservoir Caused by the Discharge of Heated Water from Browns Ferry Nuclear Plant During Open Cycle Operation," TVA Water Resources Report, 1975.
http://dudleybenton.altervista.org/pub/ML428.pdf
[3] Gerhar D Jirka, G. D. and Harleman, D. R. F., "The Mechanics Of Submerged Multiport Diffusers For Buoyant Discharges in Shallow Water," Parsons Laboratory MIT Report No. 169, 1973. http://dudleybenton.altervista.org/pub/73801.pdf

on constant discharge flow (1470 cfs/unit), constant condenser rise of 25°F, and continuous three-unit operation. The curves correspond to constant upstream temperature (i.e., 35°F, 40°F, 45°F, etc.). Above 40,000 cfs (cubic feet per second river flow rate), the curves collapse, indicating a momentum-dominated mixing process. The curves fan out at lower flows, approaching a buoyancy-dominated process. The dark blue curve (labeled 40°F) arises from the point of maximum density for water, which occurs at 39.2°F (4°C). The 35°F and 45°F curves are on top of each other, as these lie almost equidistant from the point of maximum density.

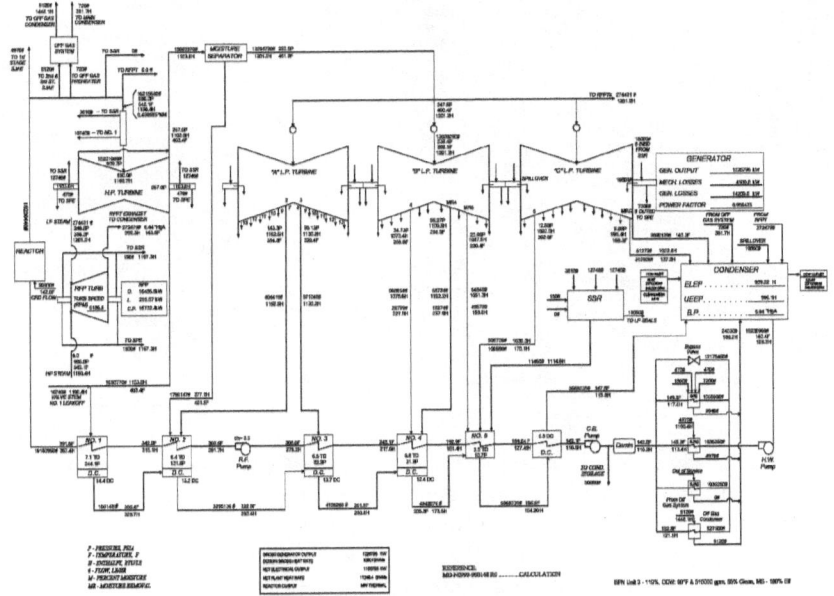

Typical Heat Balance

A system power up-rate was investigated in 2015, along with the potential impacts on the environment. The thermal aspects of these changes are documented in a summary report: http://dudleybenton.altervista.org/pub/BFTDM.pdf

Chapter 2. Wheeler Reservoir

Wheeler Reservoir is located on the Tennessee River in northern Alabama between Rogersville and Huntsville. The lake begins at Guntersville Dam and extends 60 miles to where it ends at Wheeler Dam. It is Alabama's second largest lake at 67,100 acres and having 1,027 miles of shoreline. Wheeler is one of nine reservoirs that create a stairway of navigable water on the Tennessee River from Knoxville, Tennessee, to Paducah, Kentucky.

Construction of Wheeler Dam began in 1933 and was completed in 1936. It was the second dam TVA built, finished only seven months after Norris. Wheeler Dam is 72 feet high and 6,342 feet across. The hydroelectric facility has 11 generating units with a maximum capacity of 411 megawatts. The reservoir has a flood-storage capacity of 326,500 acre-feet. To maintain the water depth required for navigation, the minimum winter elevation for the reservoir is 550.5 feet. The typical summer operating range is between 555 and 556 feet. Wheeler has two locks, one 110-by-600 feet and the other 60-by-360 feet, which lift and lower barges as much as 52 feet.

MacCormack Scheme

The one-dimensional transient partial differential equation (Saint Venant) governing open channel flow can be written in two parts: 1) conservation of mass and 2) conservation of momentum.

$$B\frac{\partial H}{\partial t} + \frac{\partial Q}{\partial x} - q = 0 \qquad (2.1)$$

$$\frac{1}{A}\frac{\partial Q}{\partial t}+\frac{1}{A}\frac{\partial}{\partial x}\left(\frac{Q^2}{A}\right)+\frac{qV_X}{A}+g\left(\frac{\partial H}{\partial x}+S_F\right)=0 \qquad (2.2)$$

where A is the cross-sectional area, B is the width of channel at the water surface, g is the acceleration of gravity, Q is the volumetric flow rate, q is the local volume inflow per unit time per unit length of channel, S_F is the slope of the energy grade line, V_X is the X-component of the velocity of the local inflow, , t is time, and x is the distance along the channel. The Manning equation gives the slope of the energy grade line:

$$S_F=\frac{Q|Q|n^2}{\left(1.486AR^{\frac{2}{3}}\right)^2} \qquad (2.3)$$

where R is the hydraulic radius (i.e., area/wetted perimeter), and n is Manning's factor. Waldrop's contribution to this process, as described in the previously referenced report, was to implement the MacCormack difference scheme, which he brought over from the aerospace industry. It had long been known that using central (finite) differences to solve Equation 2.2 required very small time steps to avoid artificial sloshing. The MacCormack difference scheme uses alternating upstream and downstream differences, which squash the unwanted solution artifacts. Details can be found in the online archive in folder examples\Wheeler along with the code (Wheeler.c). The alternating differences are accomplished with array index **new**, which is either 0 or 1 and is updated each time with **new=1-new**. The differencing core is:

```
void Routing() /* MacCormack method */
  {
  int i;
  double bbar,phpx,pqpx,pqqapx;
  for(i=0;i<=ns;i++)
    qqa[i]=sq(q[new][i])/a[i];
  for(i=0;i<ns;i++)
    {
    pqpx=(q[new][i+1]-q[new][i])/dx;
    bbar=(b[i]+b[i+1])/2.;
    phpt[new][i]=(qin[i]/dx-pqpx)/bbar;
    }
  for(i=1;i<ns;i++)
    {
    phpx=(h[new][i]-h[new][i-1])/dx;
    pqqapx=(qqa[i+1]-qqa[i-1])/(dx+dx);
    pqpt[new][i]=-gravity*a[i]*(phpx+sf[i])-pqqapx;
    }
  }
```

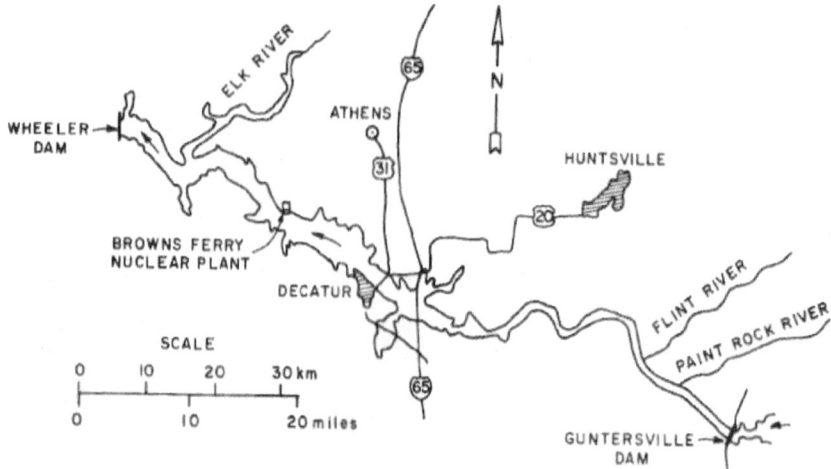

The channel geometry must be updated with each time step to account for the local elevations. The channel geometry is derived from surveying the area before the dams were built and the area was flooded.

```
void Geometry(double twup,double hwdn,double qup,double
    qdn,int new)
{
int i,i1,i2,j,j1,k,l;
double hbar,r23t,x,xj;
for(l=0;l<2;l++)
    {
    for(k=i=0;i<ns+1;i++)
        {
        i1=max(1,i);
        i2=min(ns,i+1);
        hbar=(h[new][i1-1]+h[new][i2-1])/2.;
        x=(hbar-525.)/10.;
        if(x<1.||x>5.)
            k++;
        j=max(1,min(ne-1,(int)x))-1;
        j1=j+1;
        xj=x-j1;
        a[i]=area[i][j]+(area[i][j1]-area[i][j])* xj;
        b[i]=width[i][j]+(width[i][j1]-width[i][j])*xj;
        r23t=r23[i][j]+(r23[i][j1]-r23[i][j])*xj;
        s[i]=sq(manning[i]/(1.486*a[i]*r23t));
        }
    if(!k)
        return;
    Initialize(twup,hwdn,qup,qdn);
    }
```

14

```
Abort(__LINE__,"reservior elevations have become
  unstable");
}
```

The flow routing model is set up to read the same input file as the Browns Ferry Nuclear Plant model, so that two sets of files are not necessary. Typical calculated surface elevations are shown in this first figure:

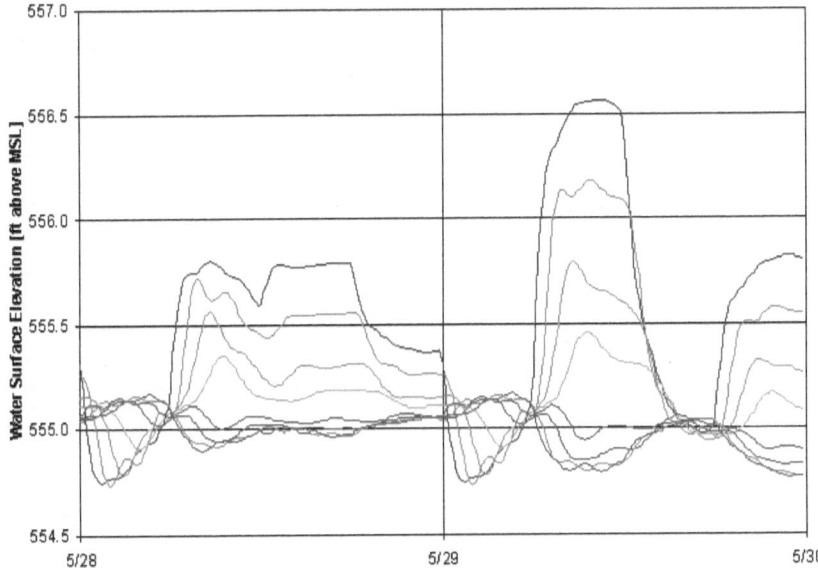

The different curves correspond to the 8 cross-sections along the channel. When considering these elevations, note that the horizontal span (left to right) is 60 miles and the vertical span is only 2.5 feet. Still, turning the dams on and off results in waves, which travel up and down the river several times before damping out.

Typical flows are shown in this next figure:

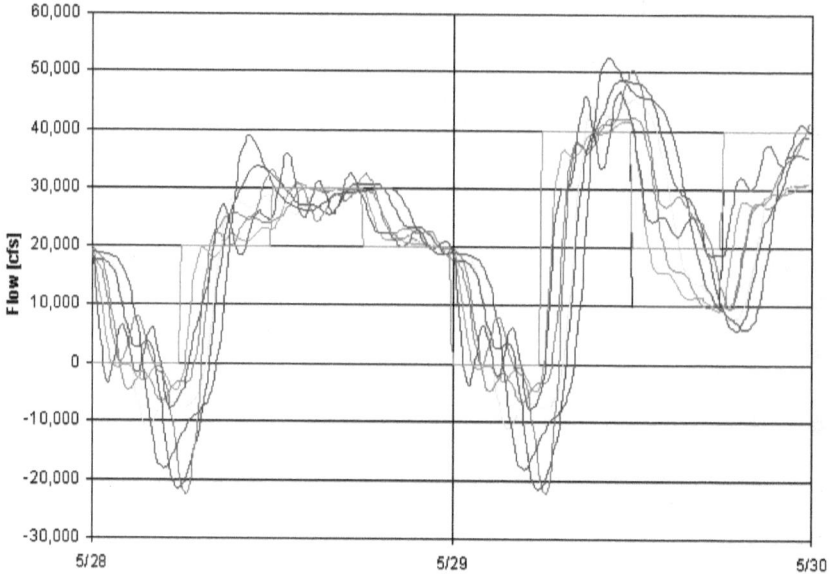

Note that the waves in the preceding figure correspond to sloshing in this figure, including significant negative flow rates along the channel. Of course, there can never be a negative flow at the dam on either end.

Chapter 3. Paradise Coal Plant

I have performed many analyses related to TVA's Paradise Coal Plant, located near Drakesboro, Kentucky. The plant has two nominally 600 MW and one 950 MW supercritical units. Units 1 and 2 were completed in 1963 and Unit 3 in 1969. Paradise is TVA's only coal-fired facility with cooling towers. The three natural draft tower, designed by Marcel LeFevre of Hamon in Belgium, were the first such towers built in North America—and the only ones to meet or exceed the expected thermal performance.

The cooling towers were a retrofit, added after Units 1 and 2 because the Green River was deemed incapable of receiving the full waste heat load of all three units in any but the highest flow season. Unlike Wheeler Reservoir, the Green River is free flowing (i.e., not controlled by an upstream and downstream dam); thus, flow routing is unnecessary. The flow can be readily estimated by the stage (i.e., water surface elevation) and is monitored by the U.S. Geological Survey (USGS).

The plant cooling system is unique and can be operated in several ways. The innovative retrofit design, which had to fit into the existing system and allow construction without shutting down the plant, was the work of Charles F. Bowman.[4] The unique design necessitates versatile software to model the cooling system, which is illustrated in the following schematic.

---

[4] Several publications are available at: http://cba-inc.com/html/publications.htm

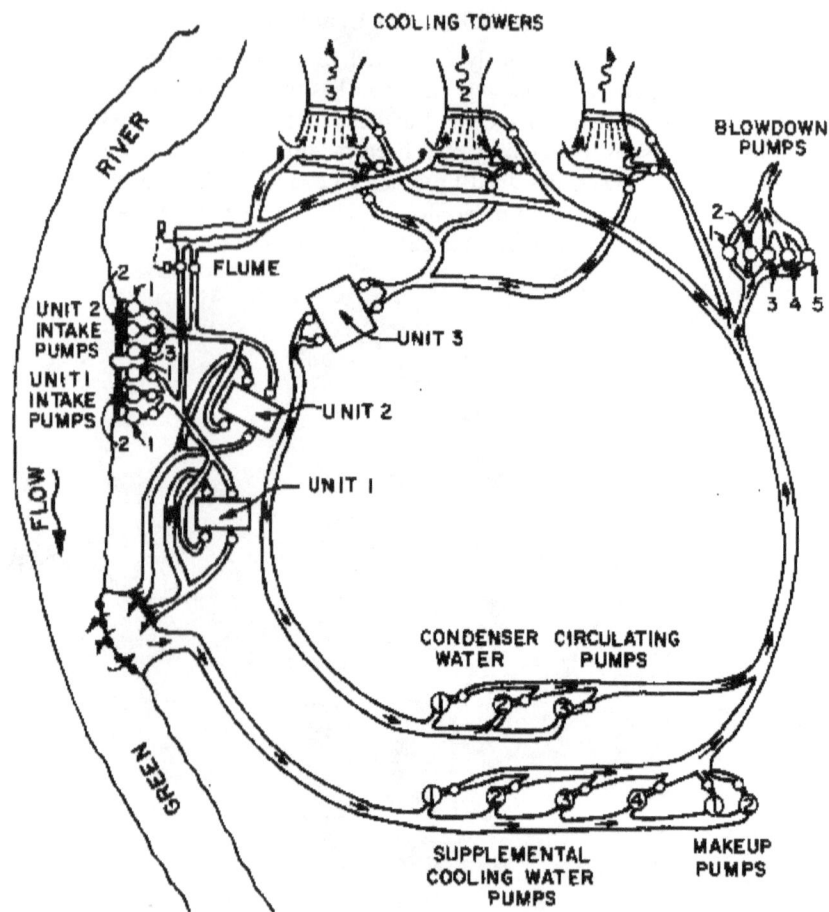

We will need two different heat load and rejection calculations (one for the two smaller units and a second for the large unit), natural draft (rather than mechanical draft) cooling tower performance (as was the case for Browns Ferry). This also means that we will need both wet-bulb and dry-bulb temperatures to drive the calculations. The river temperatures and flows were provided by the USGS. The backpressure correction curves (supercritical coal-fired) are also different from the previous (nuclear) ones, but are also a fan of curves provided by the manufacturer, also General Electric, in this case, though a different division.

The environmental regulations are also different for this plant and vary from month-to-month, requiring further complexity for the software development. Rather than simply looking at the impact on the river, with the Paradise model, the plant is presumed to reduce production as needed to meet the imposed requirements. Typical results are shown in this complex figure:

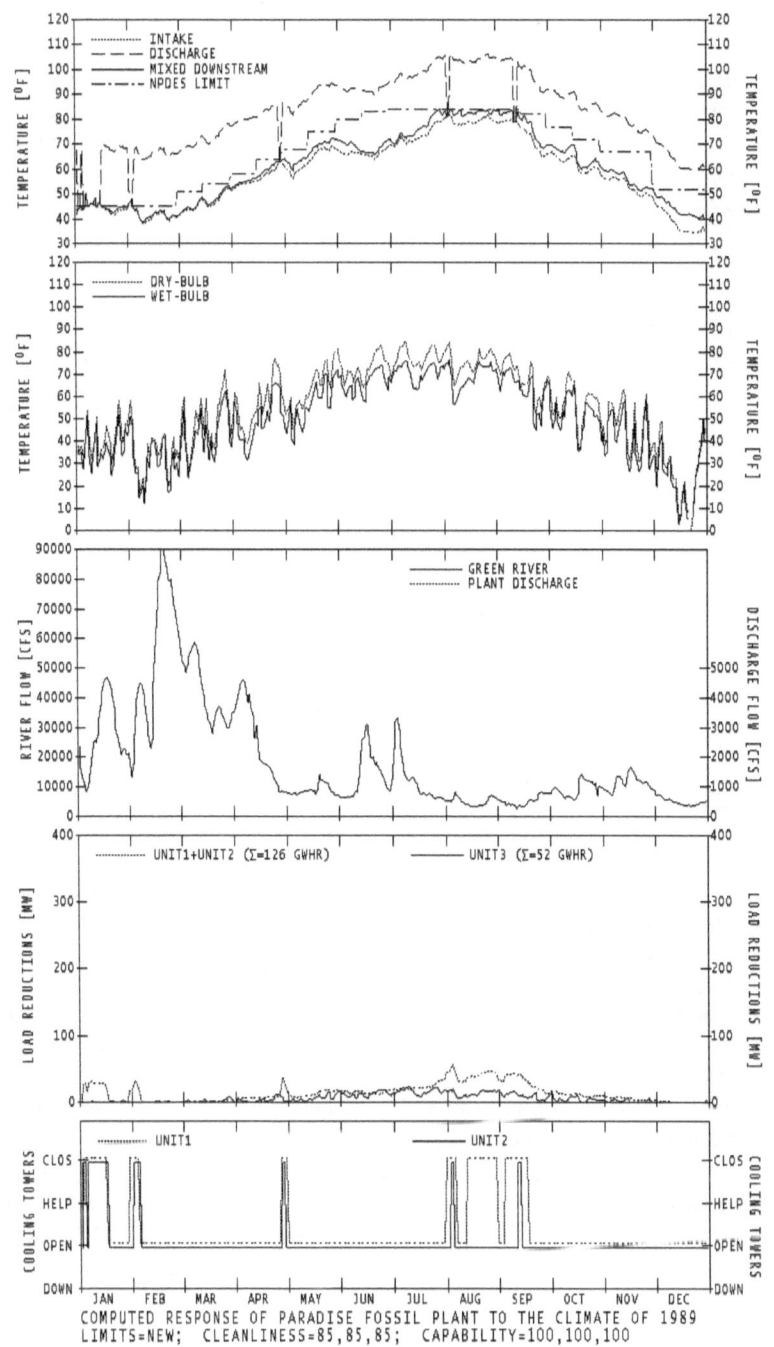

COMPUTED RESPONSE OF PARADISE FOSSIL PLANT TO THE CLIMATE OF 1989
LIMITS=NEW; CLEANLINESS=85,85,85; CAPABILITY=100,100,100

The top portion shows the measured and calculated river temperatures along with the monthly limitations. The second portion shows the meteorology (i.e., dry-bulb and wet-bulb temperatures). The third portion shows the river flows, which are seasonal and always positive. The fourth section shows the capacity (i.e., megawatt-hours of generation) lost by meeting the regulations. The bottom section shows the operational mode.

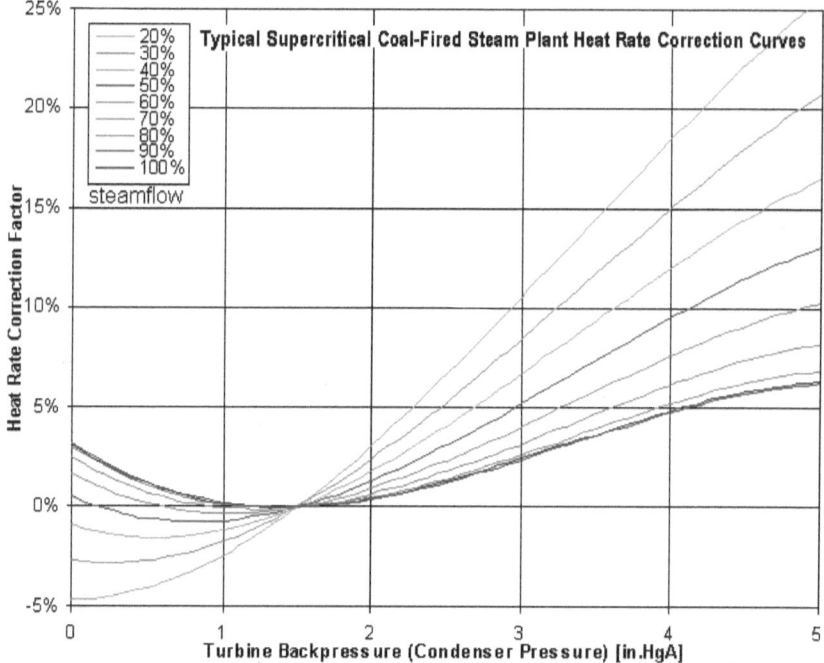

The operational mode can be one of: open, helper, or closed. These terms are common in power plant literature. Open more refers once-through water cooling without cooling tower, which is only possible for Units 1 and 2, as Unit 3 is always dissipating waste heat through one or more cooling towers. Helper mode refers to once-through water cooling with cooling tower operation, discharging the water exiting the tower to the river. In helper mode, most, but not all, of the waste heat is dissipated through the cooling tower to the atmosphere. Closed mode refers to recirculating cooling, rejecting essentially all of the waste heat through the cooling tower(s).

In closed mode, only a small portion of the cooling water is discharged (called *blowdown*) to allow for the addition of clean water (called *makeup*) to regulate the concentration of dissolved and suspended solids, which would accumulate in the system due to evaporation. Approximately 2.5% of the cooling water evaporates, making this allowance a necessity.

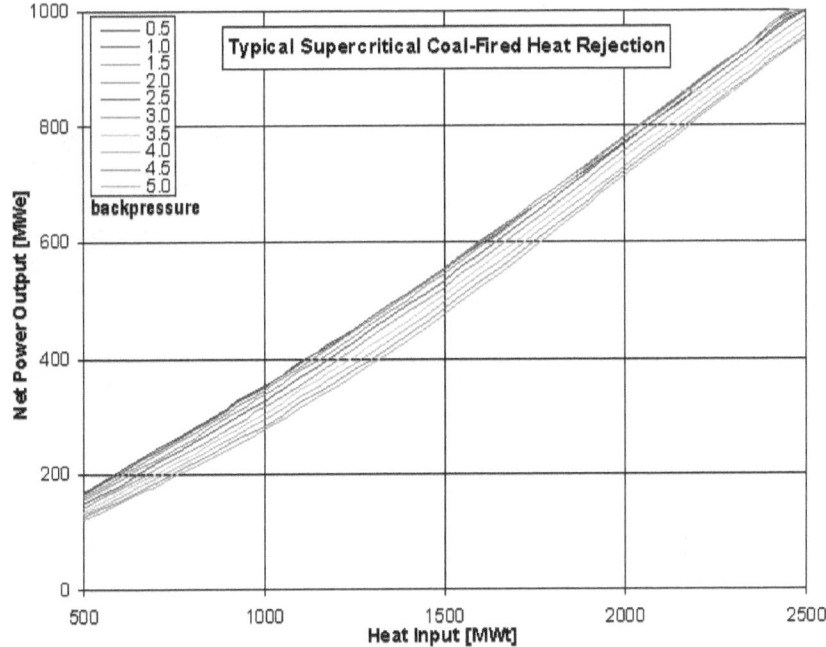

Typical Supercritical Coal-Fired Heat Rejection

Legend (backpressure): 0.5, 1.0, 1.5, 2.0, 2.5, 3.0, 3.5, 4.0, 4.5, 5.0

Y-axis: Net Power Output [MWe]
X-axis: Heat Input [MWt]

These two sets of curves indicate the amount of waste heat that must be rejected: megawatts thermal (MWt) input minus megawatts electrical (MWe) output. If this waste heat isn't dissipated into the environment, then it will be dissipated into the river. It is important to note here that the waste heat *must* go somewhere.[5]

As the environmental regulations in the State of Kentucky and on the Green River in particular changed over the course of my tenure at TVA, I was tasked with considering several operational and construction options. One way of looking at the system, including the plant and the river as the destination of any waste heat not dissipated into the atmosphere, is the thermal receiving capacity, that is, how much heat can the river receive at any time, given the conditions (flow and temperature) and regulations under consideration. The following graphic illustrates these calculations. The associated code (pafheat.c) is in the online archive in the examples\Paradise folder.

---

[5] Despite what you might have heard or read, the maximum efficiency of any thermal power plant is no more than about 45%, which means the remaining 55% will be absorbed by the environment one way or the other. It is simply *not possible* to build a 100% efficient engine. If you don't get this, you need to look up "Carnot Efficiency" on the web and read up on it. It's not that everyone who designs power plants is stupid, greedy, lazy, and doesn't care about the environment. Such is an inaccurate caricature painted by people who have never studied science and lack qualifications.

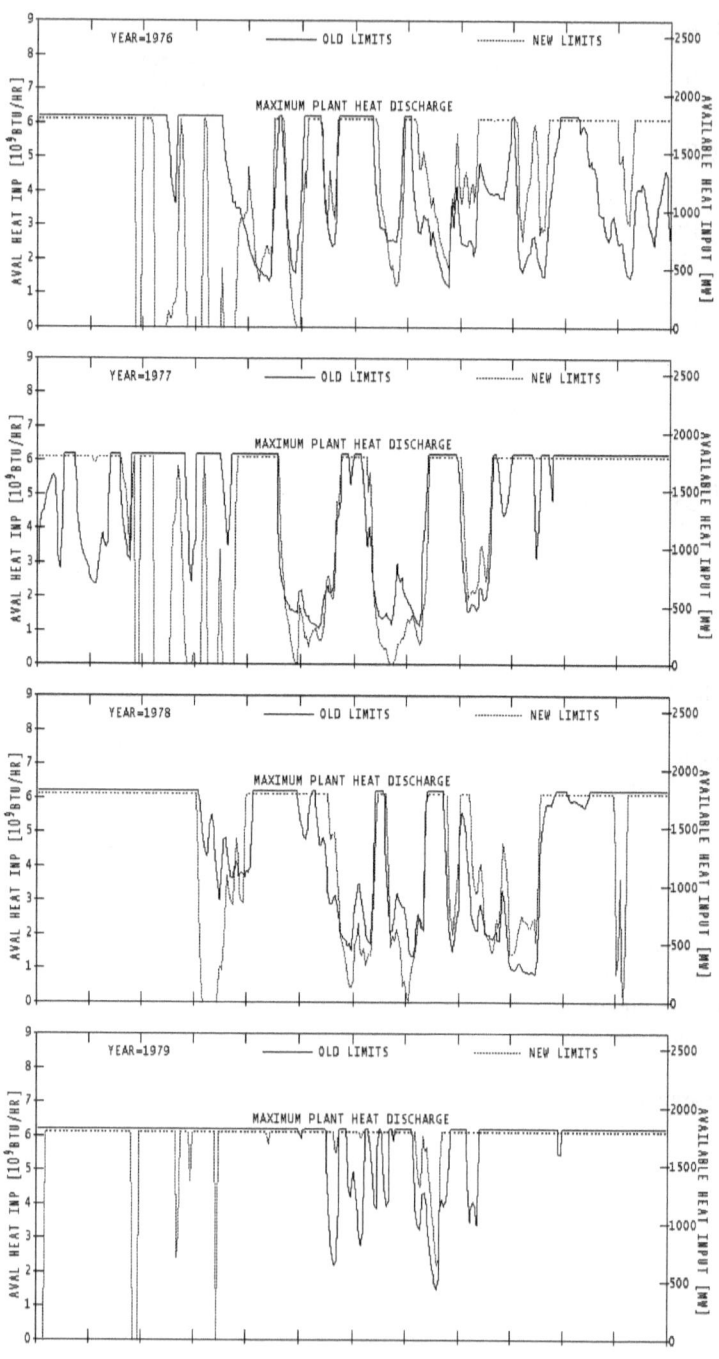

Chapter 4. Chickamauga Reservoir

Chickamauga Reservoir is on the Tennessee River, bounded above by Watts Bar Dam and below by Chickamauga Dam, which was completed in 1940. The main channel is 58.9 miles (94.8 km) long, bordering Rhea, Meigs, and Hamilton Counties with 810 miles (1,303 km) of shoreline. The dam and reservoir are named after the Chickamauga Tribe of the Cherokee Nation, who originally lived in the area. Full pool for Chickamauga lake is 682 feet (208 m) above sea level. The normal operating zone is between 675 ft (206 m) and 677 ft (206 m) through the end of March, rising steadily to a summer range of 681.5 to 682.5 ft (207.7 to 208.0 m) by the middle of May. Levels vary due to weather conditions and power needs.

The routing model is very similar to that of Wheeler Reservoir (Chapter 2) and uses the same MacCormack difference scheme to solve the one-dimensional transient open channel flow equation (2.1-2.3). The geometry and operation is different. A predictor/corrector method is used. The predictor step is Euler (i.e., explicit) and the corrector step is trapezoidal (i.e., implicit). Four corrector steps have proven to be sufficient for this geometry and time step (5 per hour).

The flows and elevations must be initialized. A ramp is used for each, based on the initial values at the upstream and downstream dam. Then two days of routing are performed and the results not written to the output file. Experiments have shown two days to be adequate in this case, but that might not be true for other geometries.

The code (Chick.c) and associated files can be found in the online archive in folder examples\Chickamauga. Several of the computational sections were listed in Chapter 2. The main routing code is listed below:

```c
void Routing(double qwb[],double qck[],double
   twwb,double hwck,double tlocal,double hiwase,double
   qsq[],double hsq[])
{
int i,iday,ihour,istep,iter,ncorr=4,ndays,nstep=5;
dt=3600./nstep;
if(initl)
   {
   ndays=3;
   for(i=0;i<ns-1;i++)
      h[new][i]=hwck+(twwb-hwck)*pow((ns-i)/13.,4);
   for(i=0;i<ns;i++)
      q[new][i]=qwb[0]+(qck[0]-qwb[0])*i/(ns-1.);
   for(i=0;i<ns-1;i++)
      qin[i]=tlocal;
   qin[7]+=hiwase/dx;
   hbar[0]=h[new][0];
   for(i=2;i<=ns;i++)
      hbar[i-1]=(h[new][i-2]+h[new][i-1])/2.;
   Geometry(hbar,a,b,s);
   }
else
   ndays=1;
for(iday=1;iday<=ndays;iday++)
   {
   if(initl!=0&&iday!=ndays)
      printf("initial day %d\n",iday);
   else
      {
      initl=0;
      nday+=1;
      printf("routing day %d\n",nday);
      }
   for(ihour=0;ihour<24;ihour++)
      {
      for(istep=0;istep<nstep;istep++)
         {
         new=1-new;                                /*
switch old/new index */
         q[new][0]=qwb[ihour];                     /* set
flows at the dams */
         q[new][ns-1]=qck[ihour];
         for(i=0;i<ns-1;i++)                        /*
predictor */
            {
            h[new][i]=h[1-new][i]+dt*phpt[1-new][i]/2.;
```

```c
              h[new][i]=fmax(675.,fmin(695.,h[new][i]));
              }
        for(i=1;i<ns-1;i++)
            q[new][i]=q[1-new][i]+dt*pqpt[1-new][i]/2.;
        for(iter=0;iter<ncorr;iter++)                    /*
corrector */
              {
        for(i=1;i<ns-1;i++)
            sf[i]=s[i]*q[new][i]*fabs(q[new][i]);
        MacCormack(qin,&q[new][0],&h[new][0],
a,b,sf,&pqpt[new][0],&phpt[new][0]);
        for(i=0;i<ns-1;i++)
            {
            h[new][i]=h[1-new][i]+dt*(phpt[1-
new][i]+phpt[new][i])/2.;
              h[new][i]=fmax(675.,fmin(695.,h[new][i]));
            }
        for(i=1;i<ns-1;i++)
            q[new][i]=q[1-new][i]+dt*(pqpt[1-
new][i]+pqpt[new][i])/2.;
              }
        if(iday==ndays)
            {
            hour+=1./nstep;
            fprintf(f6,"%lG,%lG,%lG,%lG,%lG\n",
hour,qwb[ihour]/100.,
                qck[ihour]/100.,q[new][np]/100.,(h[new][np-
1]+h[new][np])/2.);
            }
          }
        qsq[ihour]=q[new][np];
        hsq[ihour]=(h[new][np-1]+h[new][np])/2.;
          }
      }
    }
```

Typical results are shown in this next figure. Notice the presence of small- and large-scale sloshing. In this particular case, small-scale sloshing occurs between the hours of 4 and 11 in the mornings and 8 and 11 (hours 20-23) in the evenings. Large scale sloshing occurs between midnight (24) and 4 in the morning. This pattern arises because the dam releases are scheduled daily. The day shift opens the gates halfway when they arrive in the morning. The night shift opens the gates fully when they arrive and closes the gates when they leave. The number of hours the gates are opened halfway or fully open is adjusted to achieve a target seasonal volume. This is typical for large hydroelectric dams, which may provide peaking power to the grid. It's a lot easier to open or close a gate than it is to fire up or bring down a big coal-fired unit. Units such as Paradise and Bull Run are called *base load* plants, as they are not designed for transient operation and achieve maximum efficiency at full capacity.

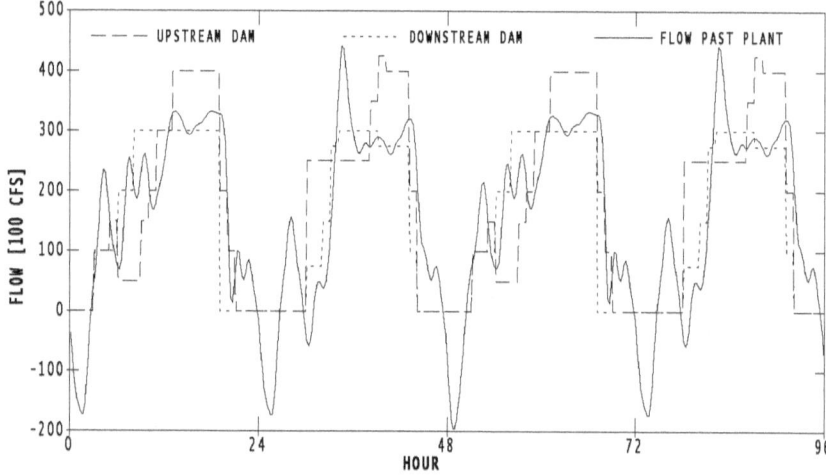

Chapter 5. Bull Run Steam Plant

Bull Run Steam Plant is located on Bull Run Creek near Oak Ridge, Tennessee. It is the only single-generator coal-fired power plant in the TVA system. When the unit went into operation in 1967, it was the largest in the world in the volume of steam produced. It also sported the latest and most expensive supercritical boiler tubes—three hundred miles of tubing. For more than a decade Bull Run was the most efficient power plant in the world.

Construction began on April 2, 1962 and was completed on June 12, 1967, when the plant began commercial operation. The nominal capacity is approximately 900 megawatts. In order to produce this much electricity, the plant burns an entire trainload of coal per day.

In spite of such auspicious beginnings, the history of this plant has been challenging. When electrostatic precipitators were added to remove particulate emissions, the extra equipment cost more than the entire plant did to begin with. The efficiency also dropped from about 43.5% to 40.5% and the plant never set a record after that. A Flue Gas Desulphurization (FGD) System (i.e., wet scrubber) was added in 2005 at a cost of a half-billion dollars, more than all the TVA coal plants cost originally.[6]

Bull Run discharges waste heat through the condensers into the Clinch River, which we will consider in more detail in the next chapter. A typical heat balance for Bull Run is shown on page iv, just after the Table of Contents. Below is a typical thermal cycle model for this plant:

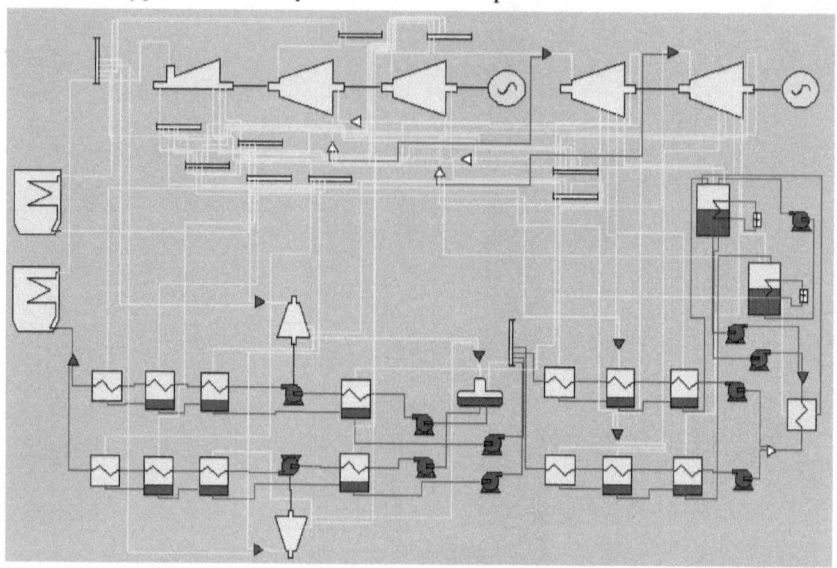

Plant performance depends on operational settings and equipment conditions as well as ambient conditions. Bull Run is an open cycle plant, having no cooling towers and no recirculating cooling water. Large coal-fired plants of this design are much less sensitive to meteorology than combined cycle plants with combustion turbines and all designs with cooling towers. Performance curves are often provided by the manufacturer, in this case General Electric. If not available, these curves can be calculated using a thermal cycle model such as GateCycle™ or QUEST, which is freely available at the web site listed in the Foreward.

---

[6] I am not arguing for a polluted environment or conditions hazardous to life and health; however, the reader should appreciate that people use a lot of electricity and it's not cheap to produce, especially if you want a clean environment too. You should also be aware that production of solar cells leaves behind various toxic substances, which may be carelessly dumped into the environment in countries where products are more cheaply produced and environmental regulations are quite lax or nonexistent.

Five sets of curves are most often used to characterize the performance of such systems. The first of these is the change in heat rate with main steam flow. The curves are for different values of the Throttle Flow Ratio (TFR), which is defined by Equation 5.1 and is described by the seminal publication on this subject by Spenser, Cotton, and Cannon, often referred to by the acronym SCC.[7]

$$TFR = \left[\frac{(flow)current}{(flow)design} \right] \sqrt{\frac{(\rho P)design}{(\rho P)current}} \qquad (5.1)$$

where ρ is the steam density and P is the steam pressure.

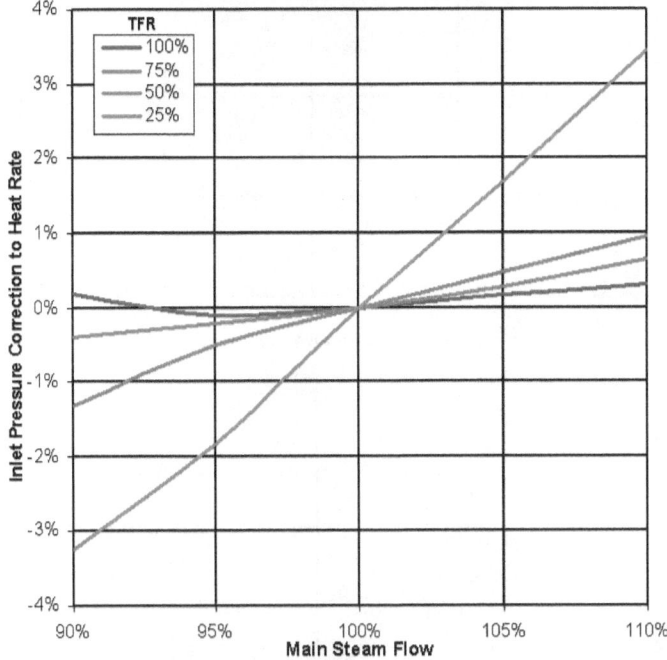

All of these curves can be found in spreadsheet BullRun.xls (and code BullRun.c) in the online archive in folder examples\BullRun. Bivariate regressions are also included as macros. Note the following relationship:

$$\Delta_{LOAD} = \frac{-\Delta_{HEATRATE}}{1 + \Delta_{HEATRATE}} \qquad (5.2)$$

[7] Spencer, R. C., Cotton, K. C., and Cannon, C. N.,"A Method for Predicting the Performance of Steam Turbine-Generators 16,500 kW and Larger," ASME Journal of Engineering for Power, Vol. 85, pp. 249–298, 1963.

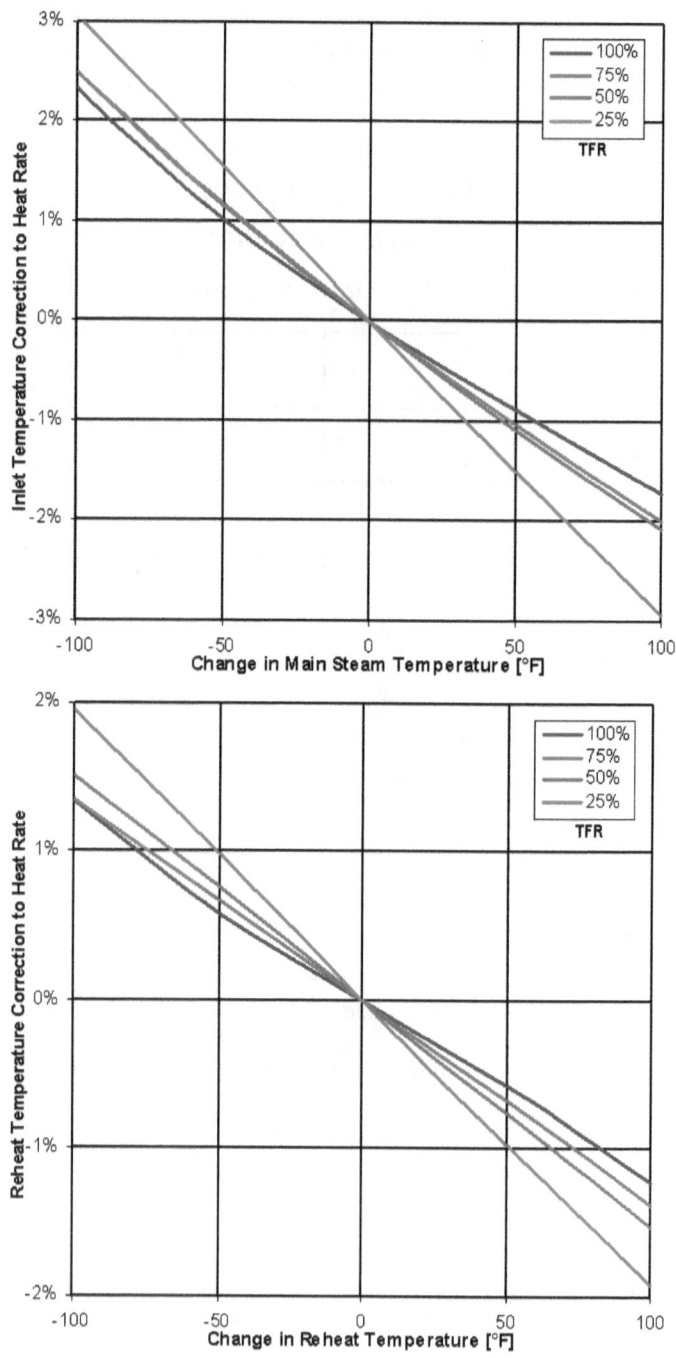

30

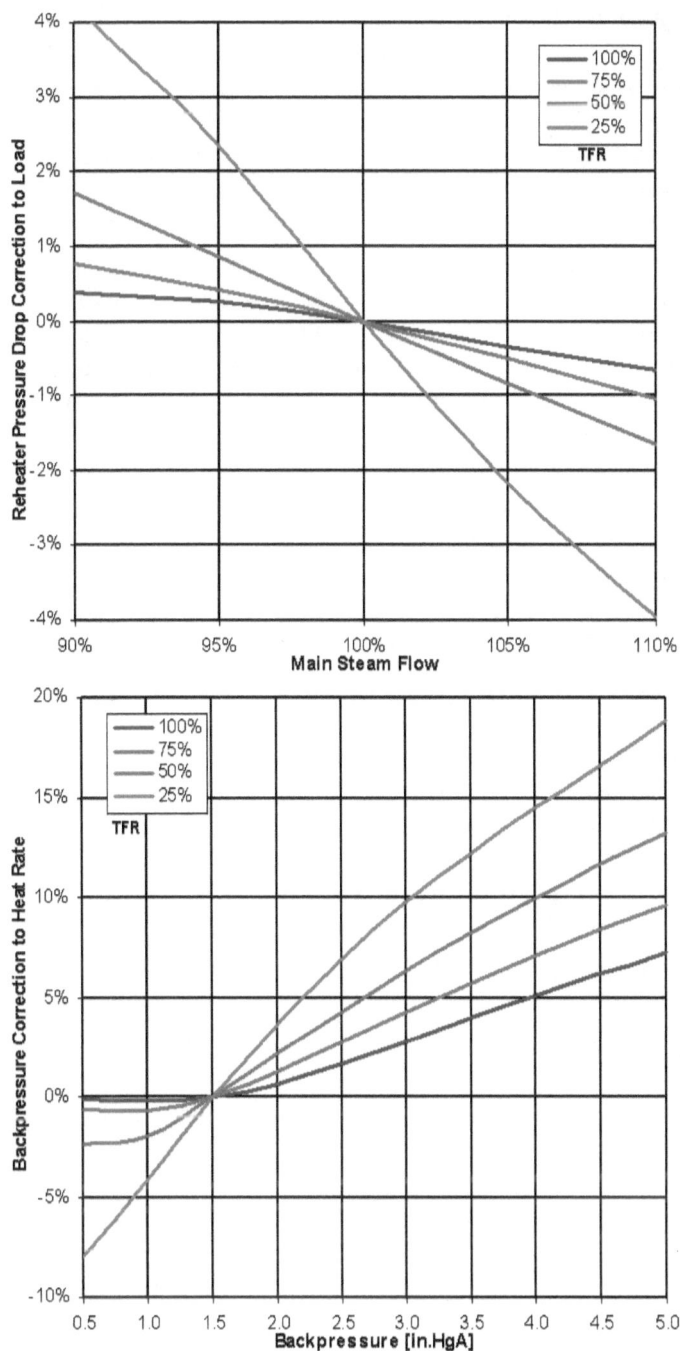

31

Output of BullRun.c is listed below:

```
              Bull Run Performance Curves
     <----- heat rate -----> <--------- load ------->
 BP  0.25  0.50  0.50  1.00  0.25  0.50  0.75  1.00
 0.5 0.055 0.042 0.026 0.006 0.058 0.044 0.027 0.006
 0.6 0.049 0.038 0.023 0.004 0.051 0.039 0.024 0.004
 0.7 0.043 0.034 0.020 0.002 0.045 0.035 0.021 0.002
 0.8 0.038 0.029 0.017 0.001 0.039 0.030 0.018 0.001
 0.9 0.032 0.025 0.014 0.001 0.033 0.026 0.015 0.001
 1.0 0.026 0.021 0.011 0.002 0.027 0.021 0.012 0.002
 1.1 0.021 0.017 0.009 0.004 0.021 0.017 0.009 0.004
 1.2 0.015 0.012 0.006 0.005 0.015 0.013 0.006 0.005
 1.3 0.010 0.008 0.003 0.007 0.010 0.008 0.003 0.007
 1.4 0.004 0.004 0.000 0.008 0.004 0.004 0.000 0.008
 1.5 0.001 0.000 0.003 0.010 0.001 0.000 0.003 0.010
 1.6 0.007 0.004 0.006 0.011 0.007 0.004 0.006 0.011
 1.7 0.013 0.009 0.009 0.013 0.012 0.009 0.009 0.012
 1.8 0.018 0.013 0.011 0.014 0.018 0.013 0.011 0.014
 1.9 0.024 0.017 0.014 0.016 0.023 0.017 0.014 0.015
 2.0 0.029 0.021 0.017 0.017 0.028 0.021 0.017 0.017
 2.5 0.057 0.042 0.031 0.024 0.054 0.040 0.030 0.024
 3.0 0.084 0.063 0.045 0.031 0.078 0.059 0.043 0.030
 3.5 0.112 0.083 0.059 0.038 0.101 0.077 0.056 0.037
 4.0 0.139 0.104 0.072 0.045 0.122 0.094 0.068 0.043
 4.5 0.166 0.124 0.086 0.052 0.143 0.110 0.079 0.049
 5.0 0.193 0.144 0.099 0.058 0.162 0.126 0.090 0.055
```

Chapter 6. Clinch River

The Clinch River is dammed twice: by Norris Dam, the first dam built by the Tennessee Valley Authority (TVA); and by the Melton Hill Dam, the only TVA dam with a navigation lock that is not located on the main channel of the Tennessee River. The Clinch River is popular spot for fishing and rowing. The section below Norris dam can be quite cold, owing to the deep withdrawal from the dam and long holdup. Before rock weirs were installed, much of the bed would periodically dry out. This particular aspect of the river is what we will consider with this next model, which I developed to inform the engineers who designed the rock weirs and were in the offices on either side of mine at the time.

The Clinch River flow routing model is different from the Wheeler and Chickamauga Reservoir routing models in that it is based on the kinematic wave equation and not the Saint Venant. It also includes a thermal model, which can handle dry bed, as this occurred regularly in the early years of operation. The files (including source code, Clinch.c, and input file, Clinch.inp) can be found in the online archive in folder examples\Clinch.

The kinematic wave equation governing open channel flow is similar to the Saint Venant equation presented in Chapter 2. We will use the same variables here as in Equations 2.1 through 2.3 to express this relationship, including the roughness and hydraulic grade line.

$$\frac{\partial Q}{\partial x}+\frac{\partial A}{\partial t}=q \tag{6.1}$$

The celerity (i.e., speed of propagation), c_K, is given by:

$$c_K = \frac{dx}{dt} = \frac{\partial Q}{\partial A} \qquad (6.2)$$

The USGS is an excellent source for documents related to kinematic wave theory, as they have been responsible for so many streams and have devoted much effort to their study.[8] There are also few TVA reports on the subject. Model variables include:

b .... channel width [ft]
c .... wave celerity [ft/s]
cbed...... specific heat of the river bed material [BTU/lbm/°F]
dbed ..... effective depth of the river bed [ft]
drybed .. dry bed indicator (last section)
dx .... length of a section [ft]
dt .... time step [sec]
edifft..... eddy thermal diffusivity [ft²/sec]
ein .... energy transfer into a section [BTU]
enew..... energy after a time step [BTU]
eold ...... energy before a time step [BTU]
eout ...... energy transfer out of a section [BTU]
h .... depth of river section [ft]
htcbed... river to river bed heat transfer coefficient [BTU/ft²/hr/°F]
htcsur ... surface heat transfer coefficient [BTU/ft²/hr/°F]
kbed ..... thermal conductivity of the river bed [BTU/ft²/hr/°F]
kwater .. thermal conductivity of water [BTU/ft²/hr/°F]
Q .... flow in a section [ft³/sec]
qin .... flow into the upper section [ft³/sec]
Qinitl.... initialization flow [ft³/sec]
Qnew.... flow in a section after a time step [ft³/sec]
Qold ..... flow in a section before a time step [ft³/sec]
Qout ..... flow out of the lower section [ft³/sec]
rbedht... heat transfer to a section from the river bed [BTU/sec]
rho........ density of water [lbm/ft³]
rhobed .. density of the rive bed material [lbm/ft³]
rn.......... Manning's number for a section
s............ friction slope in a section
s0......... bottom slope in a section
surfht.... heat transfer to a section from the atmosphere [BTU/sec]
To......... effective atmospheric temperature [°F]
Tbed..... temperature of the river bed in a section [°F]
Thad..... net thermal advection into a section [BTU]

---

[8] Miller, J. E., "Basic Concepts of Kinematic-Wave Models," USGS Report No. 1302, 1984. https://pubs.usgs.gov/pp/1302/report.pdf

Tin........ temperature of water entering the upper section [°F]
Tmax.... maximum temperature in a section as prescribed by the 2LoT[9]
Tmin .... minimum temperature in a section as prescribed by the 2LoT
Tout...... temperature of water leaving the lower section [°F]
Triv ...... temperature of a section of the river [°F]

At the time (1987), I had many years of recorded data, including release temperatures at Norris Dam and intake temperatures downstream at Bull Run steam plant. I adjusted each of the variables (dbed, effective depth of the river bed; edifft, eddy thermal diffusivity; htcbed, river to river bed heat transfer coefficient; htcsur, surface heat transfer coefficient; kbed, thermal conductivity of the river bed) to best match observations. I no longer have the data, or I would have included the calibration here. The various coefficients can be found in Clinch.c. Typical results are shown below:

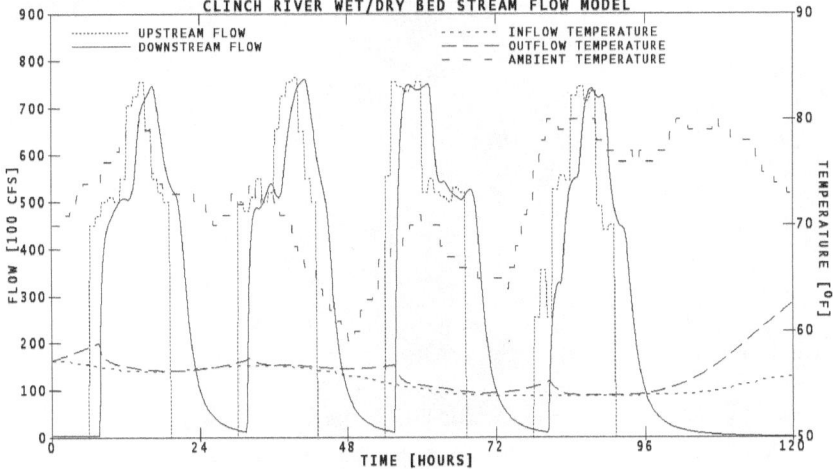

Notice that the bed dries out at about hour 100, at which time the temperature begins to rise (dashed line), reaching about 60°F at hour 120. Note also that the rise is approximately linear in this time frame, while the behavior is more like an exponential decay at other times, whether heating or cooling. There is a logical switch (drybed) in the code that controls the heat transfer calculations. As the Clinch River provides the condenser cooling water for Bull Run, this model can be combined with the one from the previous chapter to create a coupled system model similar to the one for the Paradise plant.

---

[9] the Second Law of Thermodynamics

Chapter 7. Watts Bar Nuclear Plant

I have performed many analyses for TVA's troubled Watts Bar Nuclear Plant. These challenges have arisen for a variety of reasons. Each is an interesting story in itself. Many engineers have worked to overcome these obstacles so that the plant is currently operating but has cost far more than anyone could have imagined when plans for TVA's nuclear *fleet* were first considered in the late 1960s and early 1970s. Yes, nuclear power was going to be cheap, clean, and trouble-free... not!

The figure above is a screen shot of the main window of a program I wrote to simulate the heat rejection system of Watts Bar. It is an interactive Windows® program with very few features, except that it will display curves for the reactor, steam system, cooling tower, and condenser. The reactor curves are shown on the following page. The primary use for this software is to batch process data. While it has a GUI, it also has a console (i.e., text interface), which is not interactive.

If you launch the program, providing an input file, it will simulate the thermal heat rejection system and write the results to an output file that can be pulled into excel. There are several ways to launch a Windows® file and provide an input file: 1) open a console and type in the program and file names, 2) drop the input file onto the executable using Windows Explorer® (not to be confused with Internet Explorer®, which is a web browser), or 3) create a desktop icon for the executable and supply the input file name as part of the desktop icon properties.

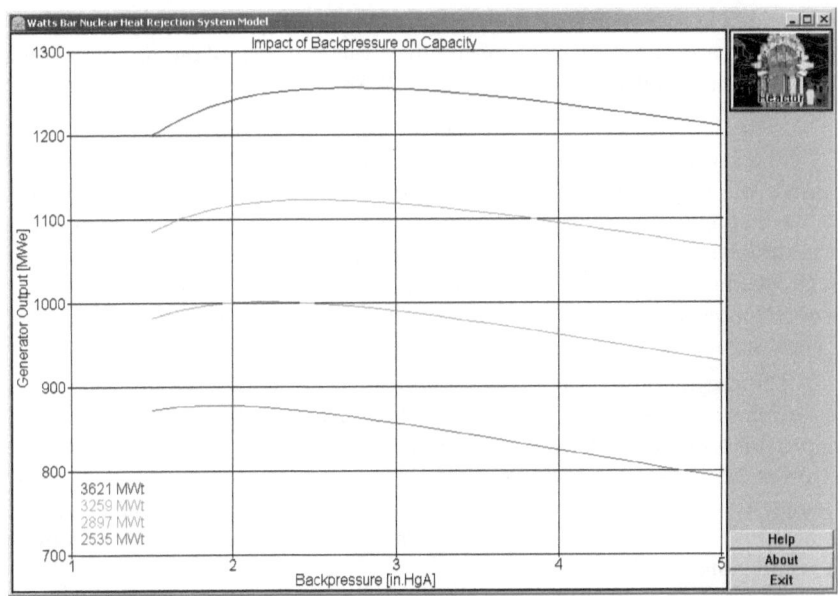

I have written 10 such computer programs that function in the same way, plant type (owner): Bellefonte Nuclear (TVA), Braidwood Nuclear (Exelon), Browns Ferry Nuclear (TVA), Comanche Peak Nuclear (Luminant), Davis Besse Nuclear (First Energy), Oconee Nuclear (Duke Energy), Paradise Fossil (TVA), Sequoyah Nuclear (TVA), Watts Bar Nuclear (TVA), Wolf Creek Nuclear (self). A sample of these can be found at:

http://dudleybenton.altervista.org/projects/Power Plants/PowerPlants.html

We will only consider four more of these in this text: Braidwood, Comanche Peak, and Oconee (with cooling lake/pond, Chapter 8) and Wolf Creek (with spray cooling, Chapter 9).

Pressurized Water Reactor

At the core of Browns Ferry Nuclear Plant's steam system is a General Electric Boiling Water Reactor (BWR); however, most of the nuclear plants built in North America are Westinghouse Pressurized Water Reactors (PWR). Watts Bar and Sequoyah Nuclear Plant steam systems are very similar. One interesting difference is that the condensers at Sequoyah are in 40-foot sections, to facilitate construction and also use standard tubing.

Due to an interesting turn of events, the condensers at Watts Bar are in a single section, having 108-foot long tubes, which must be custom-fabricated and are very difficult to transport (special railroad cars are necessary). I know of this story because friend and colleague Charles F. Bowman was responsible for making this proverbial square peg fit in the round hole of the already-in-place concrete foundation. While the BWR and PWR steam generator designs are very different, the steam turbine side of the system is quite similar so that the performance curves look very much alike, which is why these are not shown here, being so similar to the ones for Browns Ferry.

Closed-Mode Operation with Cooling Towers

The biggest difference between Browns Ferry and Watts Bar is that the former was originally designed to discharge waste heat into the river, modified to have cooling towers, but though to only need these occasionally. The latter was designed to operate with cooling towers at all times. In fact, Watts Bar has a recirculating cooling system with natural draft (i.e., tall concrete hyperbolic) towers often associated with nuclear plants, especially after the Three Mile Island Incident in 1979.

The cooling towers at Browns ferry (pictured on page 1) are mechanical draft, that is, air is forced through them by fans. Each of the fans is approximately 200 hp. Operating all of the cooling tower fans and lift pumps at Browns Ferry Nuclear Plant requires the entire output of Norris Dam. This comparison should give you an appreciation for how big nuclear plants are and how small most hydroelectric dams are in the grand scheme of power consumption.

The cooling towers at Watts Bar are natural draft. The cooling performance mechanical draft towers are dependent almost exclusively on the ambient wet-bulb temperature; however, the cooling performance of natural draft towers also depends on the dry-bulb temperature. Lower ambient relative humidity means better cooling for mechanical draft towers, but worse cooling for natural draft towers. This is because the draft (i.e., flow of air through the tower) is provided by the evaporation and heating of the air, rather than fans.

It is also important to note here that water vapor has a molecular weight of 18; whereas, air has a molecular weight of 29. The density of ideal gases is inversely proportional to the absolute temperature and the molecular weight. Water vapor is lighter than air, which is why clouds float. The more water vapor you can introduce to air without condensation, the less dense it becomes and the more buoyant, resulting in more airflow. Evaporation can also cool air, which means that the dry-bulb temperature may even drop as moisture enters the air, resulting in higher density, less buoyancy, and less airflow.

The heat load on a natural draft tower is what produces the airflow. If you decrease the heat load, you will have less airflow and less cooling. This negative feedback loop is crucial in the closed mode (i.e., recirculating) operation of Watts Bar. The same feedback does not occur when operating Browns Ferry in closed mode because the cooling towers are mechanical draft. If you drop load at Browns Ferry in closed mode in order to meet an environmental limit, the discharge temperature will drop proportionately. That doesn't happen at Watts Bar. If you drop load at Watts Bar, the natural draft diminishes, the tower produces less cooling, and the discharge temperature doesn't decrease proportionately. Under certain conditions, it requires a precipitous drop in load to achieve a significant decrease in cooling tower exit temperature—thus we have the biggest operational challenge at Watts Bar and why it matters so much that the towers did not perform as expected even when new!

We use the heat rejection curves shown in the previous figure, the heat rate corrections similar to the ones illustrated for Browns Ferry, the HEI condenser calculations described previously, the site-specific meteorology, and natural draft cooling tower performance curves, as illustrated in Appendix E to construct a unified system model for Watts Bar Nuclear Plant. The code (WattsBar.c) and associated input and output files can be found in the online archive in folder examples\WattsBar. This code also illustrates how to write a program that will run in either interactive or batch mode with a GUI.

Condenser Cooling Water

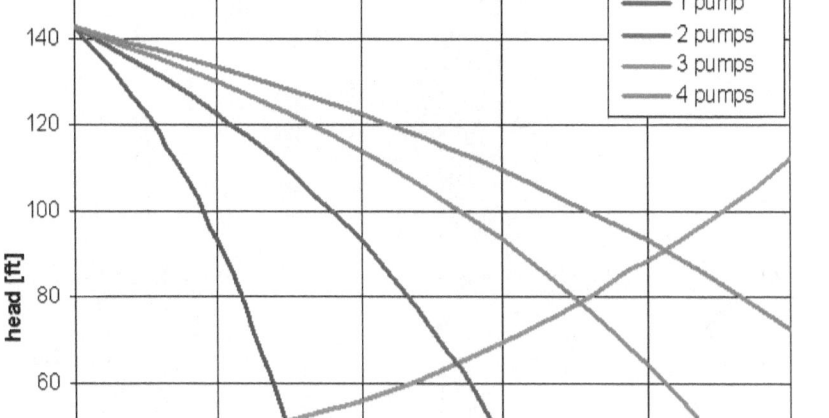

One of the most important parameters in the heat rejection system of any power plant is the condenser cooling water (CCW) flow rate. This is determined by matching the system resistance (or *head*, including friction losses and elevation difference) and the pump curves. Such plants are always designed with more than one CCW pump for several reasons, including: 1) loss of a single pump should not result in a shut down, especially for a nuclear plant, 2) the plant will not always operate at rated capacity and so full flow is not always necessary, 3) in winter months more CCW flow may actually result in lower thermal performance, in which case it is easy to just shut off one pump. The Watts Bar units are designed with four CCW pumps that also serve as cooling tower lift (CTL) pumps. Many plants (e.g., Browns Ferry) have separate CCW and CTL pumps. This balance of flow is illustrated in the preceding figure. The operating points for 1, 2, 3, and 4 pumps are at the intersections of the dark blue, light blue, dark cyan, and green curves, respectively, with the red curve.

Capacity Simulations and Scenarios

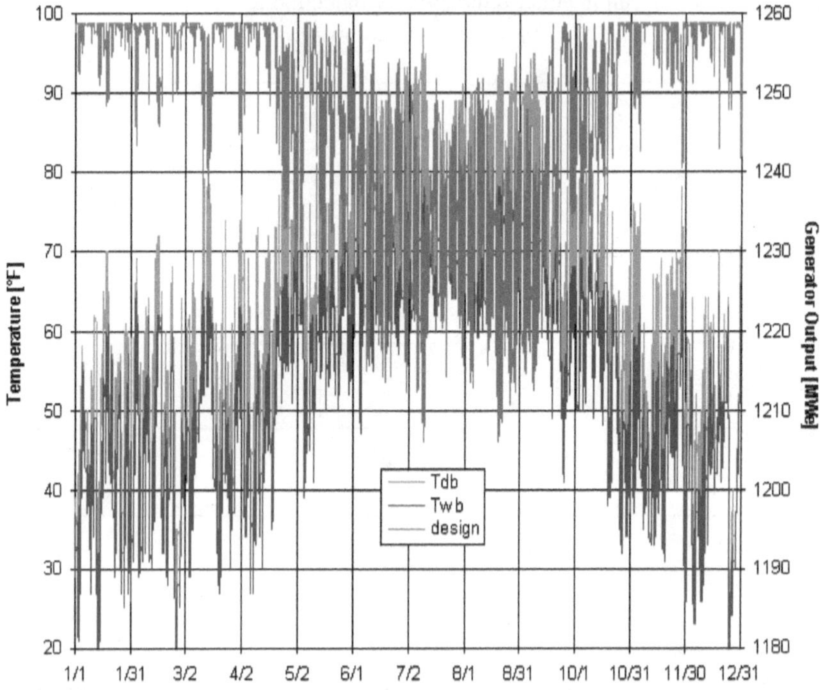

The cooling towers at Watts Bar were never quite big enough to dissipate the waste heat, which has several consequences, including: reduced generator output (and net power or capacity) and higher turbine backpressure. As there are three zones in the condensers for this particular design, each hotter than the previous one, the final (zone 3) pressure is the limiting factor. Excessive turbine backpressure (or condenser operating pressure) causes blade flutter (harmful vibrations), which result in metal fatigue and ultimately failure. In almost every steam turbine design there is a limiting backpressure, which is typically around

4.5 inches of Mercury absolute (in.HgA). Load must be reduced (reactor heat input) if necessary to avoid exceeding the maximum backpressure.

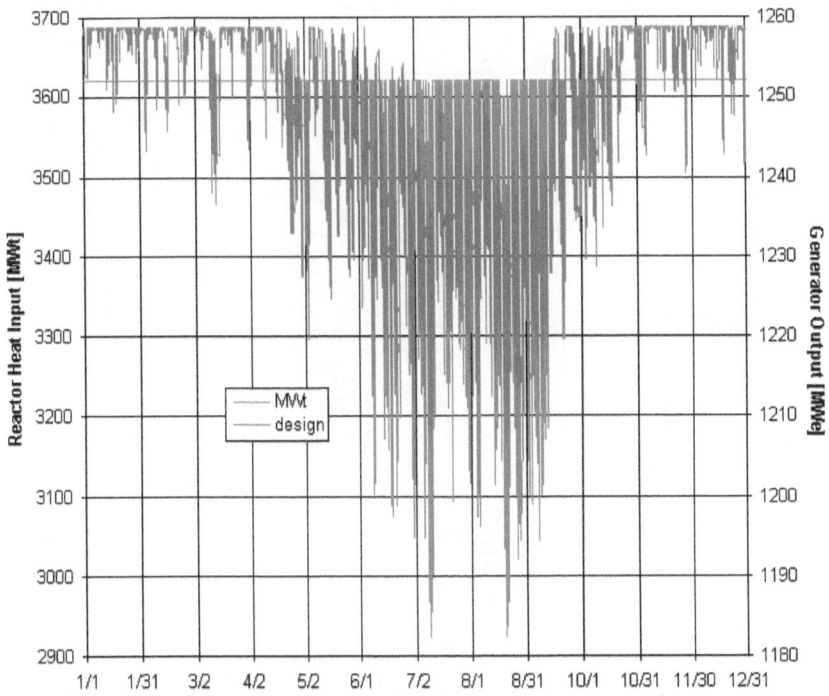

In addition to being somewhat under-designed, the cooling towers were originally tested at approximately 85% of their expected capacity. All of these variables can be changed in the code (WattsBar.c) to see the impact. I have prepared several cases to illustrate the numerous scenarios I have run with this model. In the first figure on the previous page, we see the meteorology and the expected capacity throughout the year if everything were performing as expected. In the second figure above, we see the reactor heat input (in megawatts thermal) corresponding to the generator output (in megawatts electrical). There is roughly a one-to-one correspondence. If the reactor isn't at full power, then the plant will not be operating at full capacity.

This next figure shows the zone 3 (hottest/limiting) backpressure corresponding to the estimated performance of the Unit 1 when it was first brought online in 1996. Unit 2 did not come online until 2016. A number of changes were made during the interim to address various issues and also to take advantage of new technologies and innovative strategies for cooling water intake and discharge of the blowdown.

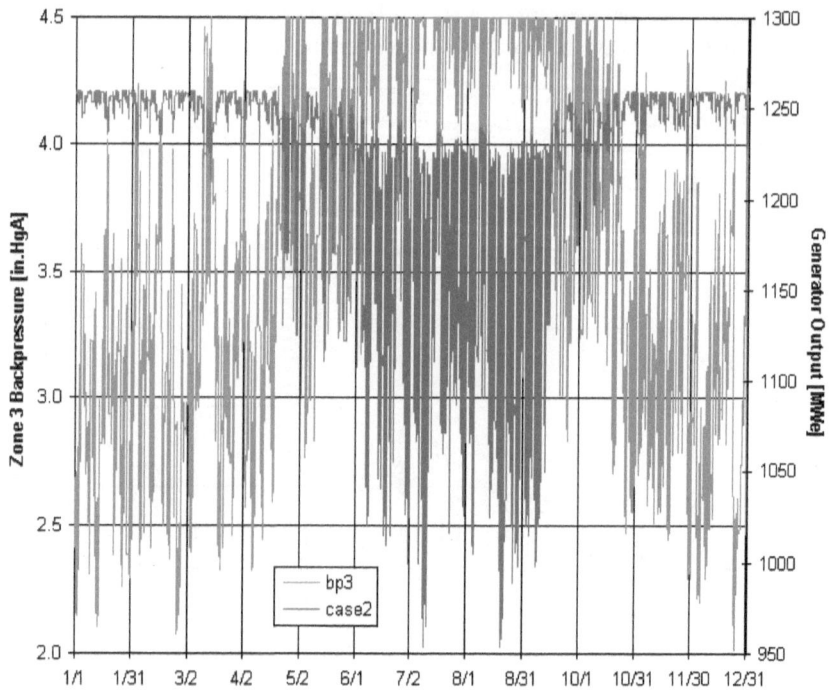

A shortfall in cooling tower performance was first identified and thought to be the limiting factor in capacity. During the initial investigation, the limiting backpressure was found to be far more controlling and greatly compounded by any under performance of the cooling tower. An additional study was quickly undertaken to see if the limit could be raised from the typical 4.5 to 5.0, 5.5, 6.0, and even 6.5 in.HgA. The manufacturer (Westinghouse) quickly agreed to 5.5, but was hesitant to provide assurances beyond that value. After much negotiation and shifting of responsibility from the manufacturer to the operator, as well as installing additional vibration monitors and an associated alarm system, a value of 6.5 in.HgA was agreed upon. That compromise, along with restoration of the design cooling tower performance, yields the baseline results in this particular simulation.

The second scenario (expected cooling tower under) performance with upper limiting backpressure is labeled, "case 1". Original (as-built) cooling tower performance plus the original (typical utility rating) backpressure is labeled, "case 2". This next figure shows the lost capacity resulting from these two scenarios.

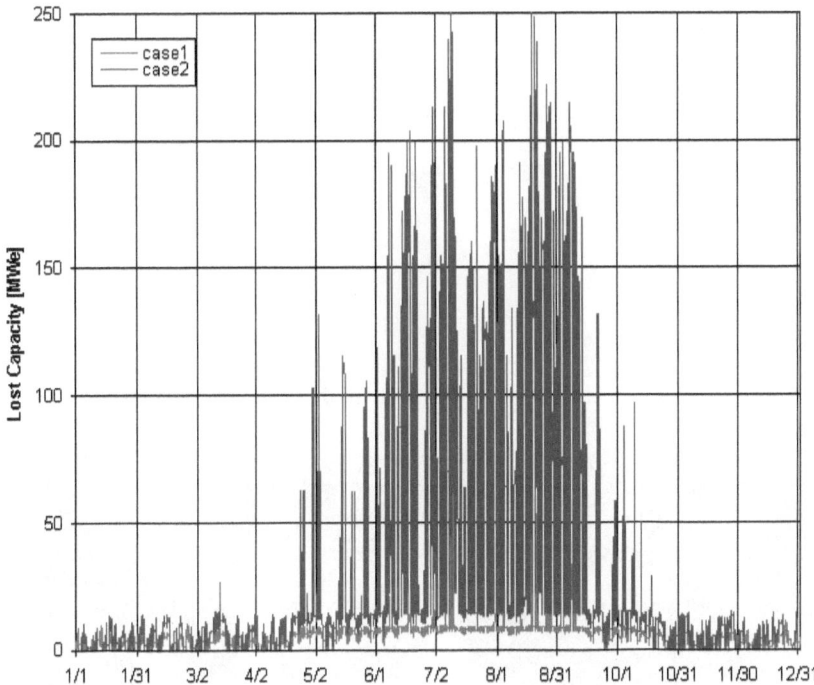

While case 1 appears to be much less severe than case 2, it still corresponds to more money lost than you and I will make in a lifetime. The fact that load-limited operation is predicted for both of these cases beginning in mid March and extending through late October is significant from an economic standpoint as well as for dispatch. In this region, running air conditioners in the summer is when utilities hope to make money, not limp along or run inefficient peakers.

The code (WattsBar.c) is easy to use and modify. The functions are setup for convenient running of scenarios. The code listed below illustrates how to implement a series of cases.

```
void Tables()
  {
  double ht;
  Di=Do-2.*Wt[gauge-12];
  printc("pumps gpm\n");
  for(Npump=1;Npump<=4;Npump++)
    printc("  %i %6.0lf\n",Npump,FindFlow());
  printc("Twb  RH  MWt  MWe  bp1  bp2  bp3\n");
  Twb=55.;
  rh=0.71;
  for(ht=2400.;ht<3800.1;ht+=100.)
    {
    heat=ht;
```

43

```
      bpmax=6.5;
      Simulation();
      }
    printc("Twb   RH   MWt   MWe   bp1   bp2   bp3\n");
    for(rh=0.2;rh<1.01;rh+=0.2)
      {
      for(Twb=30.;Twb<90.1;Twb+=10.)
        {
        heat=HEAT;
        bpmax=6.5;
        Simulation();
        }
      }
    }
```

The program writes these tables out at the end of every batch run, as illustrated below:

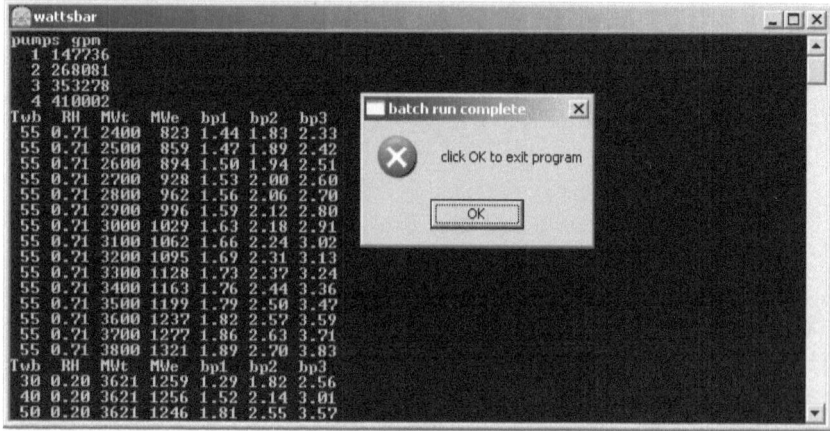

Chapter 8. Cooling Ponds and Lakes

Braidwood, Comanche Peak, and Oconee Nuclear Plants all rely on a lake or pond to dissipate waste heat, ultimately into the environment. There was much interest in large cooling ponds during the early days of nuclear power plant design and construction in the United States. The Nuclear Regulatory Commission (USNRC) and the U. S. Environmental Protection Agency (USEPA) funded a series of studies extolling the efficacy of such ponds for waste heat removal.[10,11,12,13] These articles may be found on-line.

Equilibrium Temperature

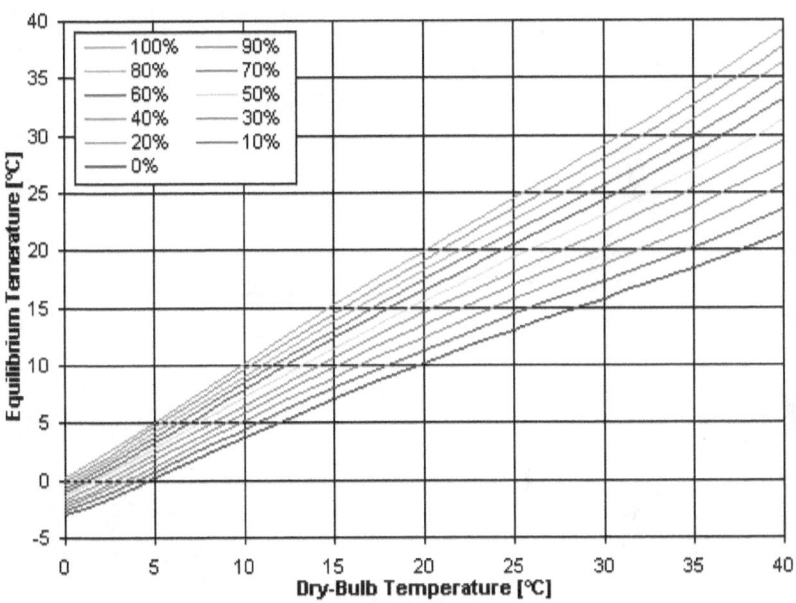

Most of these studies refer back to the work of Langhaar.[14] While this is a very old reference, the properties of air and water have not changed. Langhaar's

---

[10] Hogan, W. T., Liepins, A. A., and Reed, F. E., "An Engineering - Economic Study of Cooling Pond Performance," EPA Project 16130DFX05/70 Contract No. 14-12-521, May, 1970.

[11] Hadlock, R. K. and Abbey, O. B., "Thermal Performance Measurements on Ultimate Heat Sinks - Cooling Ponds," NUREG/CR-0008, February 1978, also published as a Batelle Pacific Northwest Laboratories Report PNL-2463.

[12] Codell, R. and Nuttle, W. K., "Analysis of Ultimate Heat Sink Cooling Ponds," NUREG-0693, November, 1980.

[13] Berger & Taylor include some helpful discussion as well (see Reference 34).

[14] Langhaar, J. W., "Cooling Pond Many Answer Your Water Cooling Problem," Chemical Engineering, August, 1953, pp. 194-199.

analysis was quite insightful and his calculations continue to be relevant and useful. This method has been updated and refined for presentation here. The calculations have been implemented in an Excel® spreadsheet (Langhaar.xls in folder examples\Langhaar). All cooling pond calculations begin with the definition of an equilibrium temperature. This is the temperature a pond will eventually approach under steady conditions. The preceding figure is based on Langhaar's nomograph. It is important to note that the equilibrium pond temperature is not simply the web-bulb; rather, it is based on empirical data. The following figure shows the difference between the equilibrium and web-bulb temperatures:

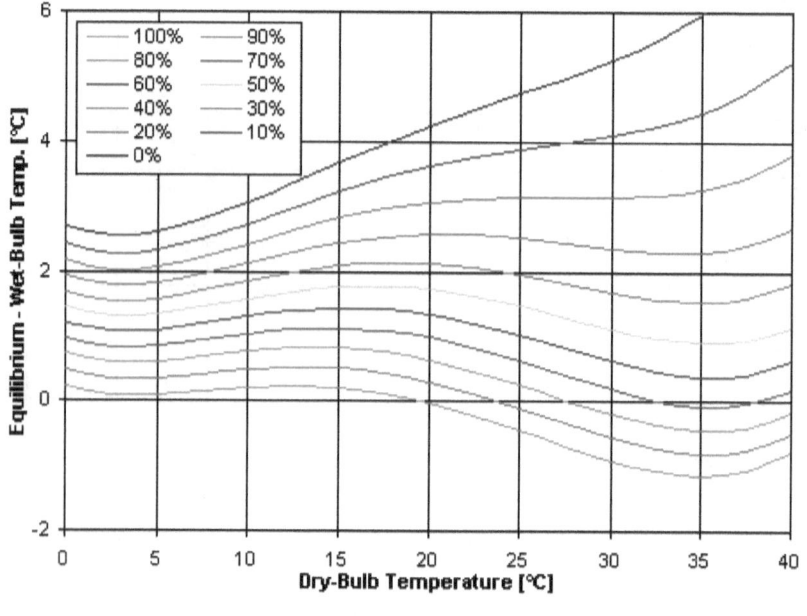

Wind and Solar

Wind and solar are strong influences on pond temperature, so that any model must contain corrections for these factors. The following figure shows the impact of wind speed on the equilibrium temperature with a Global Horizontal Irradiance (GHI) of 750 W/m². There are similar plots for 1000, 500, and 250 in the spreadsheet. The functions to perform these calculations are also included as macros. Langhaar's method includes two more parameters related to scale, given the symbols P and Q. The product $PQ=A/F$ is the area divided by flow and is dimensional so that unit conversions are necessary for implementation. SI units are used in these first several graphics and the spreadsheet. English units are used in the subsequent figures and code (Braidwood.c).[15]

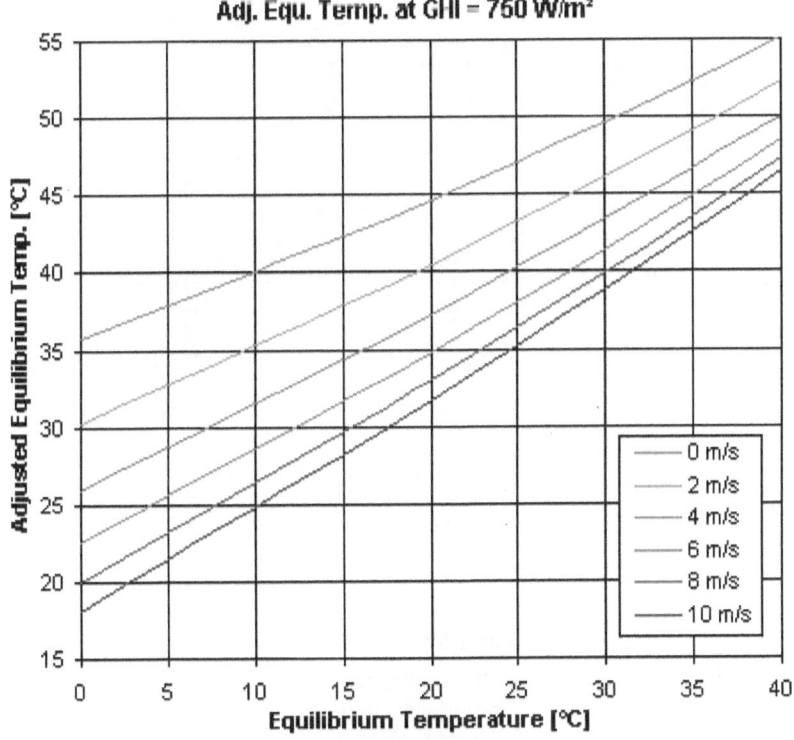

Adj. Equ. Temp. at GHI = 750 W/m²

(Y-axis: Adjusted Equilibrium Temp. [°C], X-axis: Equilibrium Temperature [°C])

Legend:
— 0 m/s
— 2 m/s
— 4 m/s
— 6 m/s
— 8 m/s
— 10 m/s

<u>Case-in-Point: Braidwood</u>

The best way to illustrate Langhaar's model of pond performance is through an actual application where a large-scale pond is used to cool a power plant, in this case nuclear. The thermal performance of the plant is needed in order to capture the response of the heat rejection to the ambient conditions. Performance functions for Braidwood Nuclear Station may be found in examples\Braidwood in Braidwood.c. This performance is typical for Westinghouse pressured water reactors. All of the files required to build the interactive Windows® application are also in this folder. The program will also run interactively or batch, as with the Watts Bar code, which is similar in other respects, except for the cooling pond instead of cooling towers.

---

[15] The practicing scientist/engineer should be comfortable working with any and all systems of units. Arguments over units are both unproductive and foolish.

Pond Geometry

Pond area is determined from bathymetric surveys. Regression is used to fit area vs. elevation. Volume is determined by integrating the area with respect to elevation from the bottom up. The pond geometry is shown in this next figure:

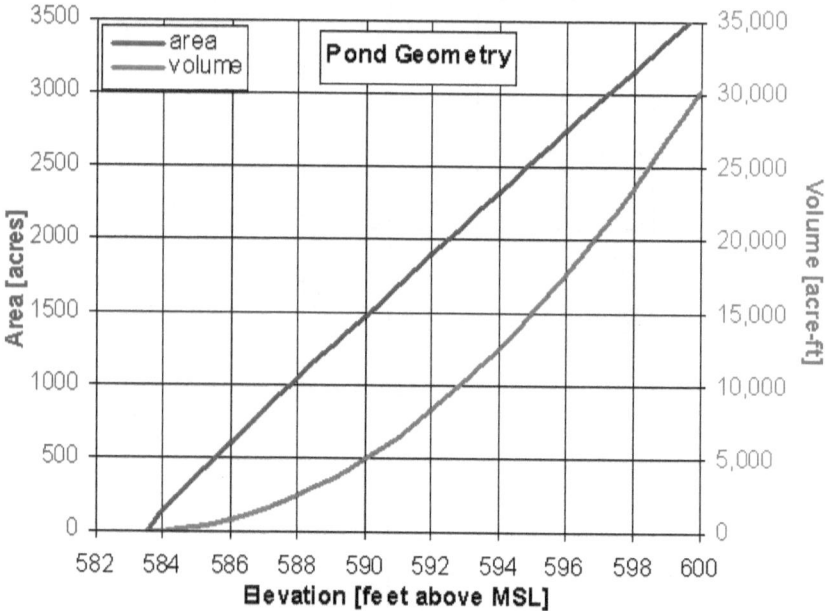

Langhaar's calculations are arranged so as to yield a required pond size for a given heat load, ambient conditions, and equilibrium temperature. The spreadsheet and code solve the inverse equation for equilibrium temperature that yields the actual pond size.

Residence Time

Depending on elevation and the number of circulating water pumps operating, the time required to circulate the entire pond volume is between 1.5 and 3.5 hours. Point-by-point simulations using the equilibrium temperature are adequate, long-term simulations should consider this as well and so it is built into the program (Braidwood.c).

Meteorological Data

In order to drive the cooling pond model we need wet-bulb, wind speed, and solar heat input. We calculate wet-bulb from barometric pressure, dry-bulb, and dew-point, which is the way the National Weather Service and the NCDC report meteorology. We estimate the net solar heat input from the Global Horizontal Irradiance (GHI), which is most often reported in W/m². Langhaar's method requires English units (i.e., BTU/hr/ft²) and also has a different base reference level. Typical meteorological data for this site is shown in the next two figures.

48

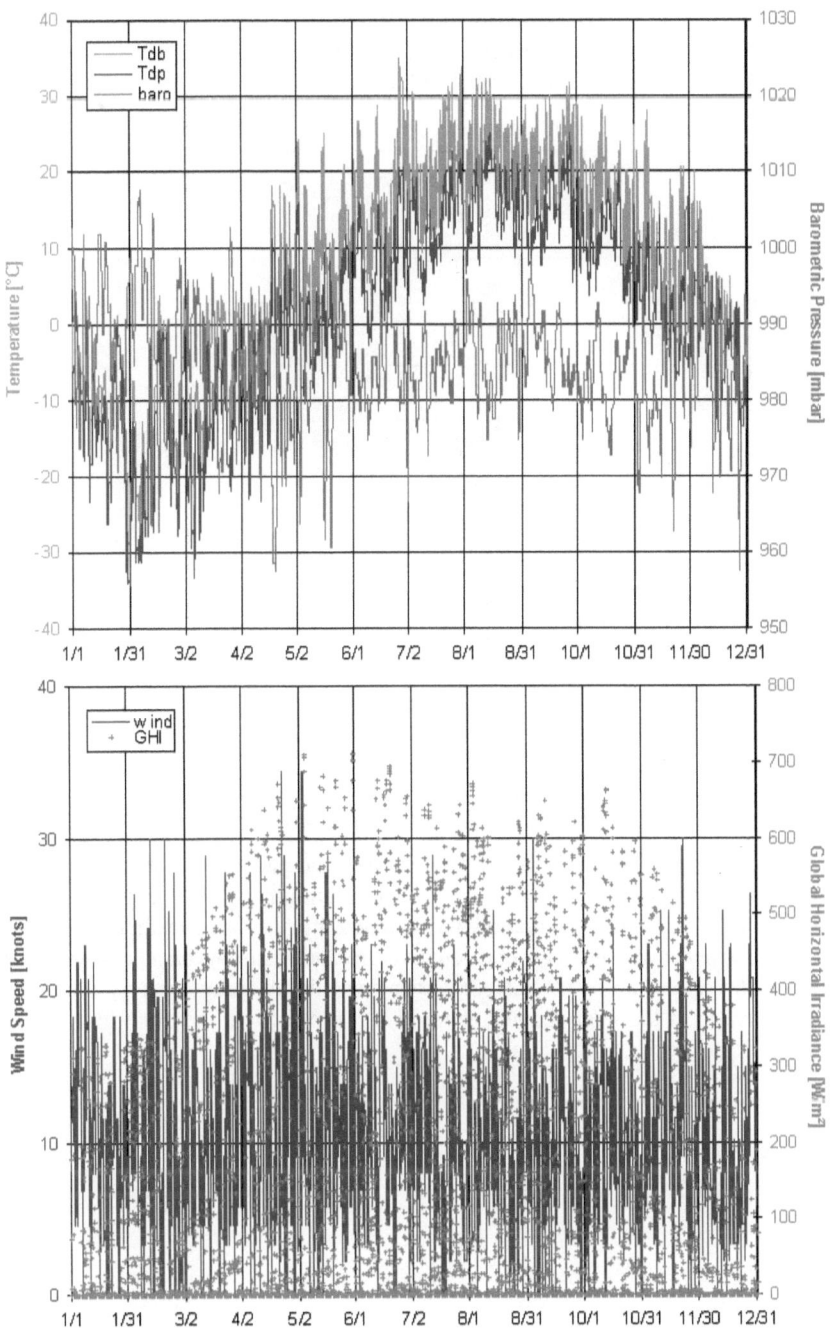

This information (file data.csv in the Braidwood folder) is dropped onto the executable (Braidwood.exe) to run the time-series simulation, which is written to the output file (Braidwood.csv). The load and backpressure are shown in this next figure:

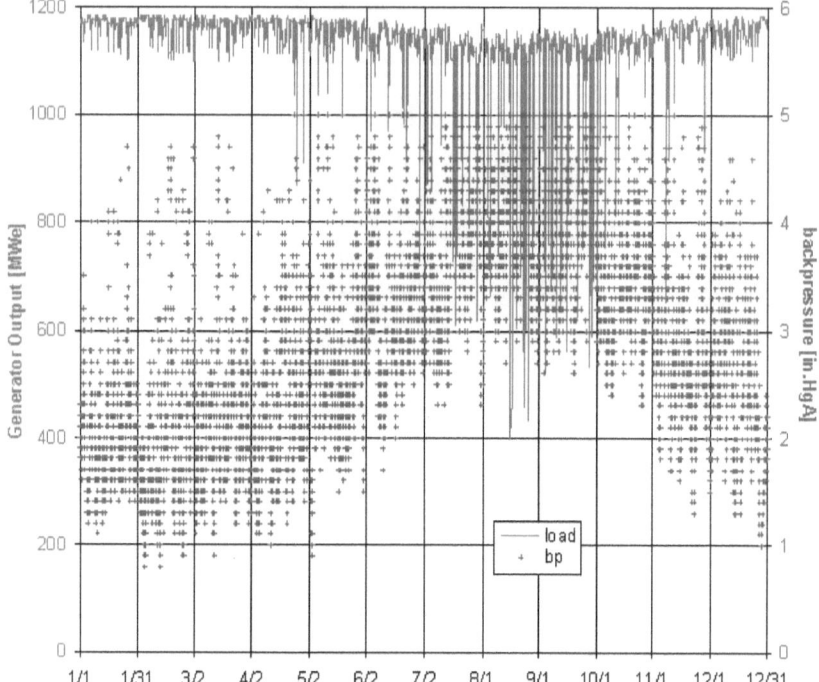

The generator output is fairly steady at near the rated level with a maximum of 1183 MWe, an average of 1144, and a minimum of 399. This last value represents a very deep reduction necessitated by the hot weather and solar heating, which drive up the pond temperature and resulting backpressure. The average backpressure is 2.9 in.HgA. Other than adding spray coolers, which we will consider in the next chapter, there isn't much that can be done to increase the cooling capacity of this lake.

The condenser inlet and exit (Tcw and Thw, respectively) in this next figure are hourly and the pond temperature (Tpond) is a three-hour rolling average to account for the residence time of the lake.

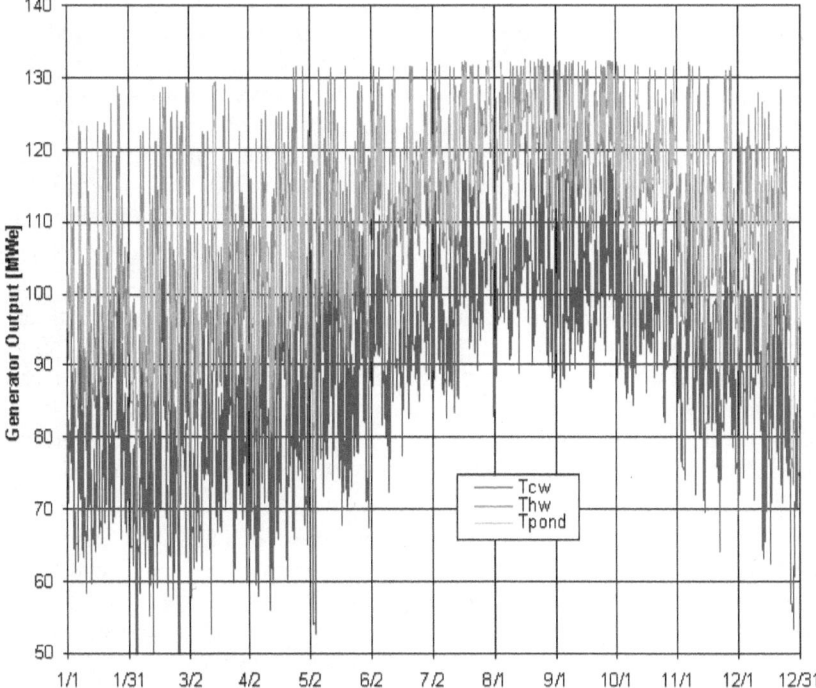

This model (Braidwood.c) is organized in such a way as to facilitate running different scenarios, such as adding spray cooling. The economics can easily be estimated by the change in generation (megawatt-hours) plus some consideration for maintaining base load operation and not having to make adjustments to avoid the backpressure limit and possible damage to the last stage steam turbine blades.

Chapter 9. Spray Cooling

Interest in large-scale spray cooling followed shortly after that of cooling lakes and ponds, as described in the previous chapter. Much of this interest was funded by the USNRC for consideration in providing ultimate heat sink (UHS) cooling. Nuclear reactions do not instantly cease when the control rods are inserted plus the reactors themselves have significant thermal mass. The steam-to-steam system in a BWR and the water-to-steam system in a PWR also have considerable thermal mass. The steam systems proper of these systems are the largest ever built, typically 1100 to 1300 MWe. Bull Run, once the largest coal-fired plant in the world, having a nominal capacity of 900 MWe, operates at a main steam flow rate of 6,350,000 lbm/hr (800 kg/s). Watts Bar, a PWR having a nominal capacity of 1100 MWe, operates at a main steam flow rate of 15,500,000 lbm/hr (1950 kg/s).

The thermal inertia of a nuclear plant is truly staggering. In the event of a forced shut down, all of this heat must be dissipated, regardless of what might have triggered the shut down—including a tornado or earthquake. This means there must be a simple, reliable (i.e., *ultimate*) heat sink always available. As spray cooling systems are quite simple (compared to a cooling tower) and hopefully won't blow away or fall down, these seemed like an ideal choice for a UHS.

There are many publications on spray ponds, but these can be readily summarized by two observations: 1) they don't perform very well and 2) the bigger they get, the worse they perform. Persistent research efforts over decades have significantly improved this gloomy outlook. The biggest improvement has been oriented spray cooling. Several publications are available on this topic as well. These are all summarized and culminated in the Power2019 article of Bowman, Taylor, and Hubble.[16] In this one source, you will find all of the relevant derivations, equations, correlations, and discussion.

Case-in-Point: Wolf Creek

Wolf Creek Nuclear is located near Burlington, Kansas. Wolf Creek, was dammed to create Coffey County Lake, which is the ultimate source of water for the plant, that has one 3565 MWt Westinghouse PWR, which began commercial operation in 1985. The unit was originally rated at 1170 MWe. A new turbine generator rotor was installed in 2011, increasing the capacity to 1250 MWe.

Arguments have surrounded the plant since it's inception, including that of water usage. The water issue became so "heated" that spray cooling was considered. This author was tasked with developing a complete heat rejection model of the plant with it's existing systems as well as spray cooling in order to

---

[16] Bowman, C. F., Taylor, R. E., and Hubble, J. D., "The Oriented Spray Cooling System for Heat Rejection and Evaporation," Proceedings of the ASME Power Conference and Nuclear Forum, 2019.

evaluate different scenarios. I completed that development in September of 1998. Ownership of the plant has also been volatile so that controlling interest has shifted over its lifetime, as recorded by official documents of the USNRC. Although I completed the work as proposed and on schedule, the decision-makers had changed and it was determined that no valid contract existed and the work was no longer needed. While this means I didn't get paid, it also means that I'm under no obligation and I never signed a confidentiality agreement. A similar story can be told of my Braidwood and Oconee projects.

And so it is that we now consider spray cooling for Wolf Creek Nuclear. Let us first consider currently working systems. Once such implementation is the spray cooling pond at the Rostov Nuclear Power Plant in Volgodonsk, Russia is shown in this next figure:

It can be challenging to quantify the cooling achieved by such systems. Fortunately, the supporting analysis has been published.[17,18] Frohwerk's patent contains an excellent description of spray systems.[19] The airflow induced by the spray process carries away the absorbed energy in the form of sensible and latent heat. The airflow and heat transfer are intimately linked. The conservation of energy can be expressed as follows:

---

[17] Berger, M. H. and Taylor, R. E., "An Atmospheric Spray Cooling Model," Proceedings of the 2nd AIAA/ASME Thermophysics and Heat Transfer Conference, Palo Alto, California, May 24-26, pp. 59-64, 1978.
[18] Chaturvedi, S. K. and R. W. Porter, "Thermal Performance of Spray-Canal Cooling Systems," Journal of Engineering for Power Vol. 102, No. 4, pp. 776-781, 1980. (available free on-line, search for 19770077431.pdf)
[19] United States Patent No. 3,622,074 issued to Paul A. Frohwerk of the Ceramic Cooling Tower Company on November 23, 1971. (available free on-line)

$$G\Delta h = LC_p\Delta T_W \qquad (9.1)$$

Lagrangian droplet tracking can be applied for a range of droplet diameter, d, and falling height, h, to produce the following graph of 1-effectiveness, or approach/(range+approach):[20]

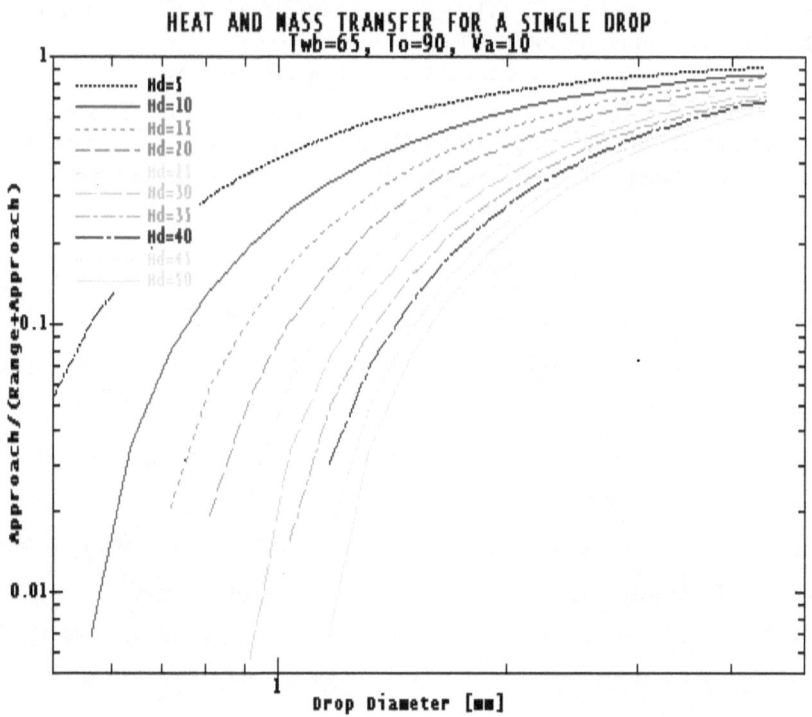

HEAT AND MASS TRANSFER FOR A SINGLE DROP
Twb=65, To=90, Va=10

Based on regression, these curves can be approximated by the following formula for $d \geq 0.5$mm:

$$\varepsilon = 1 - \left\{ e^{[a+b\ln(h)+c\ln(d)]} - 1 \right\}^3 \qquad (9.2)$$

where a=-2.456 b=0.6317, and c=-1.369. As indicated at the top of the figure, this is for a single droplet. Spray nozzles produce a distribution of droplet sizes. The following distribution of droplet sizes is typical for such nozzles. The blue dots are measured data and the red curve is a lognormal distribution having a mean of 0.43 mm and a standard deviation of 1.9 mm. The droplet size distribution can be applied to the preceding effectiveness curves to arrive at a

[20] Benton, D. J. and R. L. Rehberg, "Mass Transfer and Pressure Drop in Sprays Falling in a Freestream at Various Angles," International Association for Hydraulic Research, Fifth Cooling Tower Workshop, Palo Alto, California, 1986.

single curve of effectiveness vs. height for the droplets created by this nozzle at this operating pressure. The performance of the distribution of drop sizes is equal to the sum of the probability times the performance for each drop size, $\varepsilon=\sum P\varepsilon/\sum P$. The sum of the probabilities is equal to one.

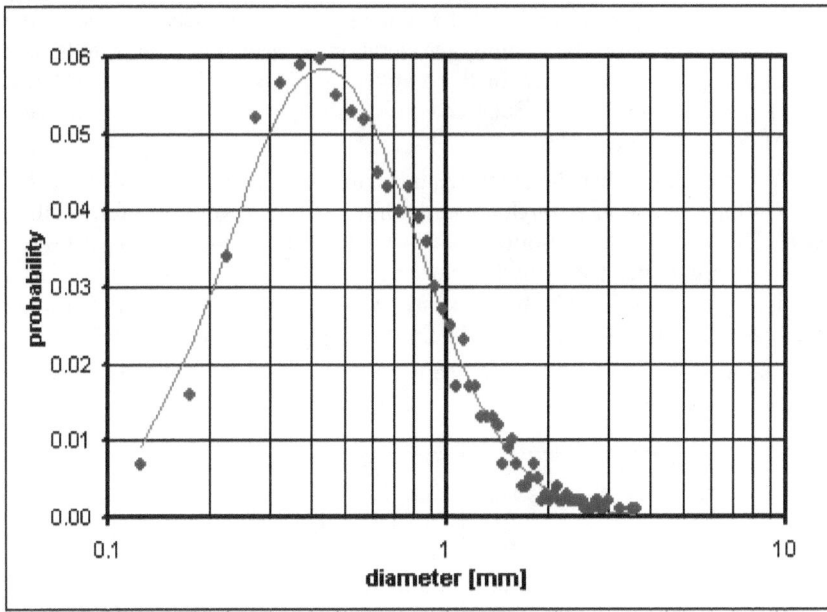

Approach to the wet-bulb is illustrated in this next figure:

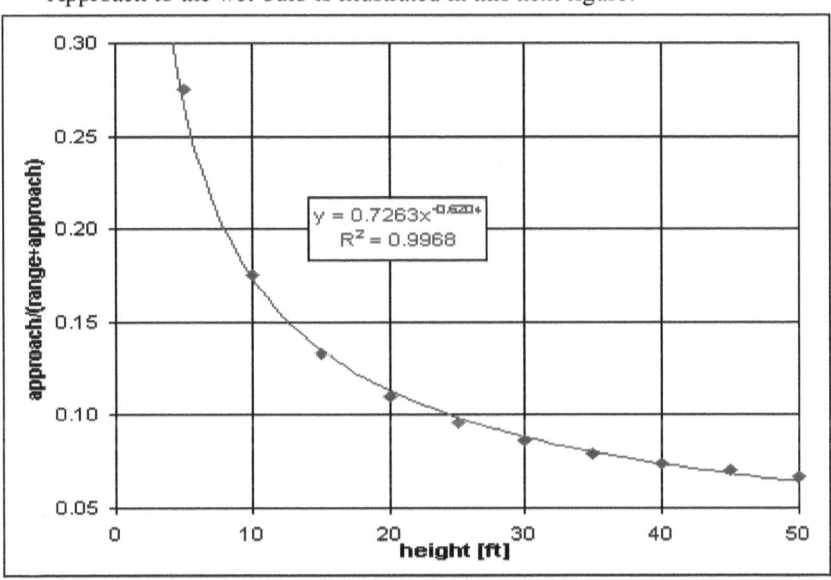

As water vapor (molecular weight ≈18) is lighter than air (molecular weight ≈29), evaporation from the spray will induce an upward draft of air, even in the absence of a crosswind. In the case of an open spray, as in the typical power plant application depicted here, the air flows radially inward and then upward.

Floating spray modules have been deployed for decades in ponds and channels for supplemental cooling and aeration at steel mills, pulp mills, and wastewater plants. Utilization of these systems at power plants has been less common, at least in the U.S. Such systems can be quite cost-effective and there are still a few manufacturers.

It is fortuitous that the same relationship for heat transfer and draft that governs the flow of air through a natural-draft cooling tower holds for a falling spray. This is particularly useful, as the problem of modeling this type of flow is much more complicated. This is primarily due to the fact that the flow in a cooling tower is confined by the structure; whereas, the flow induced by a spray is not. The fill also serves to guide the flow through the tower structure.

The following figure shows temperature data collected for the spray cooling pond at a paper mill:

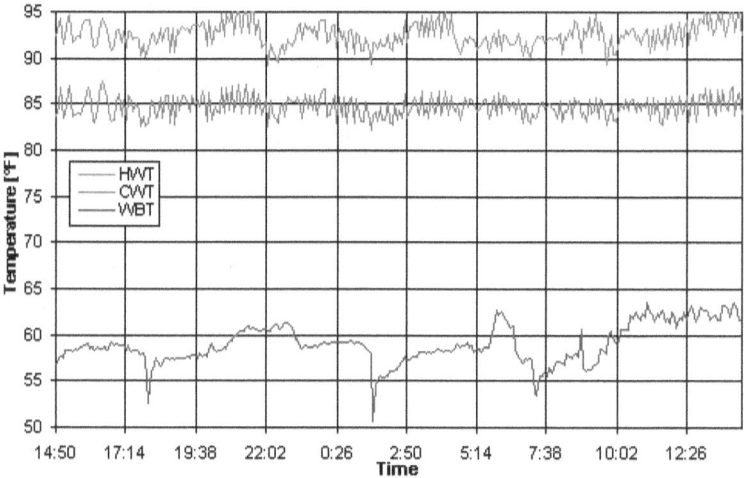

The water-to-air mass flow ratio, **L/G**, can be calculated from Equation 9.1 and the mass transfer characteristic, **KaV/L**, can be calculated from Merkel's Equation 9.3. The first figure on the next page shows the results of these calculations for this particular spray system and the second shows agreement with data over a month.

$$\frac{KaV}{L} = C_{PW} \int_{T_{OUT}}^{T_{IN}} \frac{dT}{h_F - h_A} \qquad (9.3)$$

56

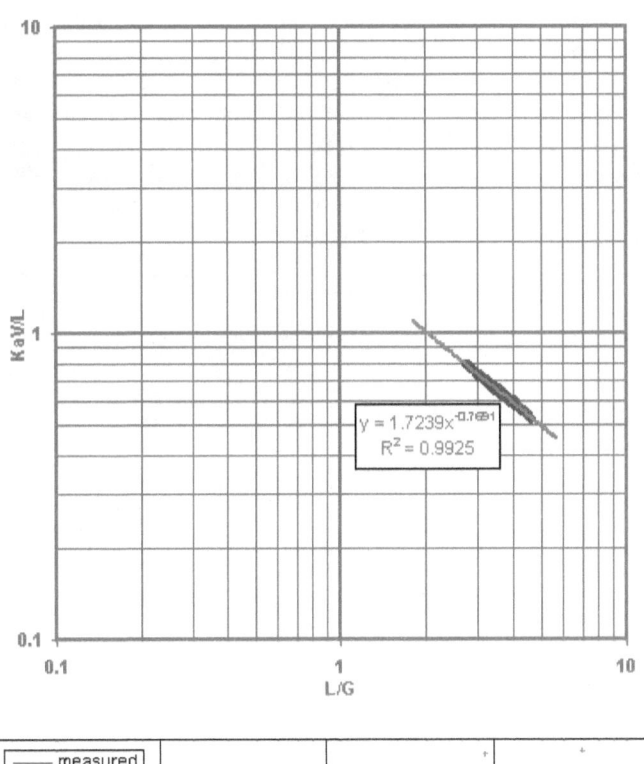

$y = 1.7239x^{-0.7691}$
$R^Z = 0.9925$

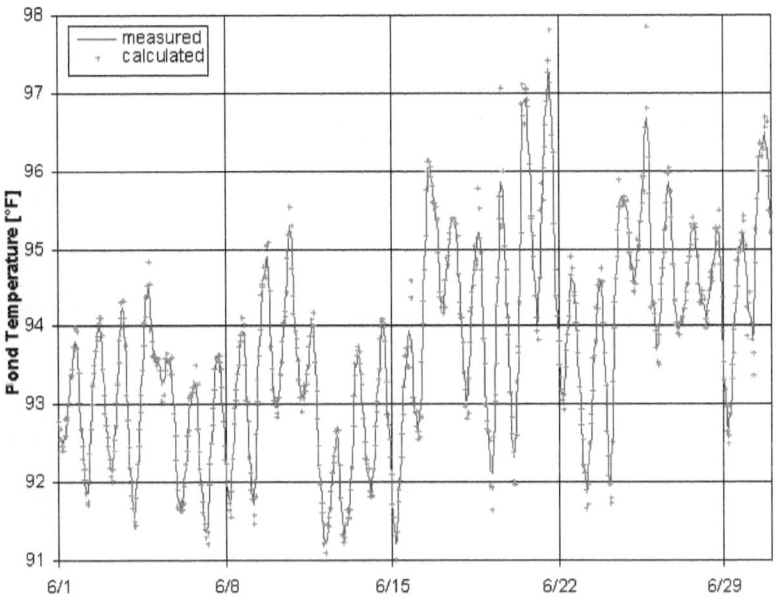

Data for this spray cooling system produces a remarkably tight cluster ($R^2=0.9925$) about the line $KaV/L=1.7239/(L/G)^{0.7691}$, clearly illustrating the facility of Merkel's theory and this approach. Of course, Wolf Creek Nuclear would require a much larger spray cooling system than this paper mill, but the same principles apply to either case. Data fro June of 1997 was provided for this study, but could easily have been pulled from the GSOD, as described in Appendix B and C. The meteorology plus expected condenser inlet and exit temperatures are shown in this next figure:

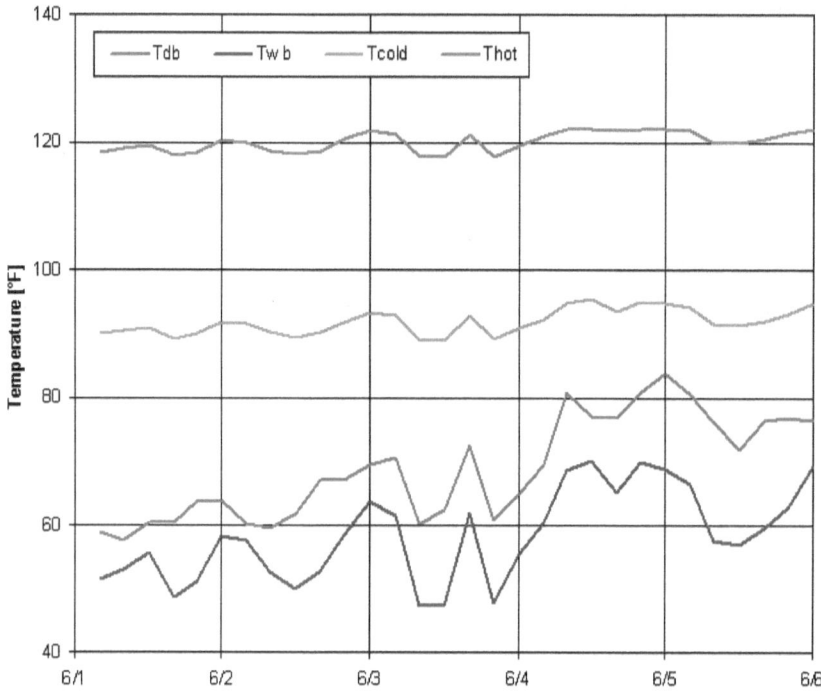

The reactor heat input and generator output are shown in this next figure. Note that a significant de-rate event (i.e., necessary drop in load to control backpressure) was predicted on June 4, 1997. This is one of the things the operator was particularly interested in when considering spray cooling.

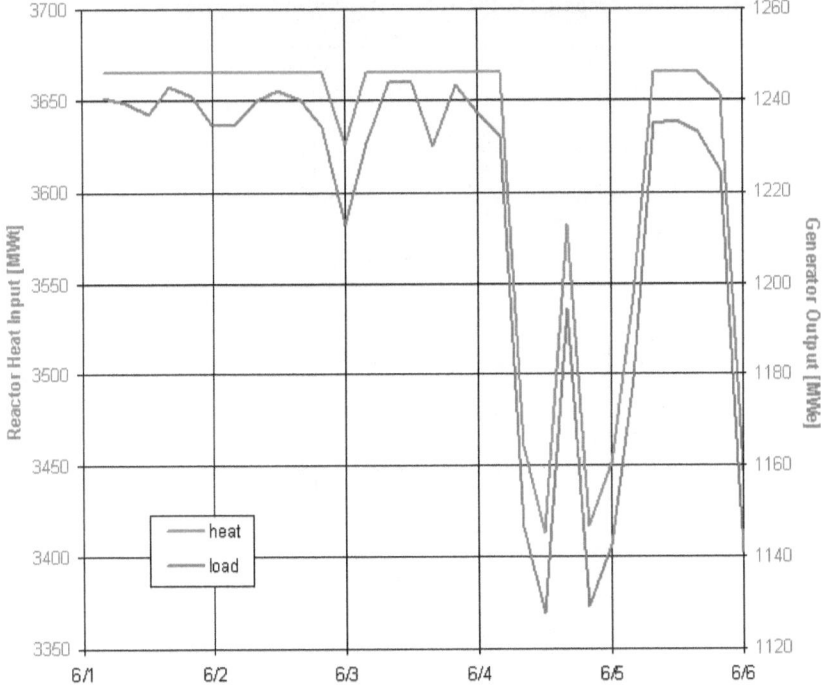

The three zone backpressures are shown in this next figure. Wolf Creek, like Watts Bar, is a Westinghouse PWR with a three-zone multi-pressure condenser. The CCW flow at Wolf Creek (500,000 gpm) is somewhat larger than for Watts Bar (410,000 gpm) and there are three separate water boxes, each with 29-foot tubes compared to the 108-foot tubes at Watts Bar.

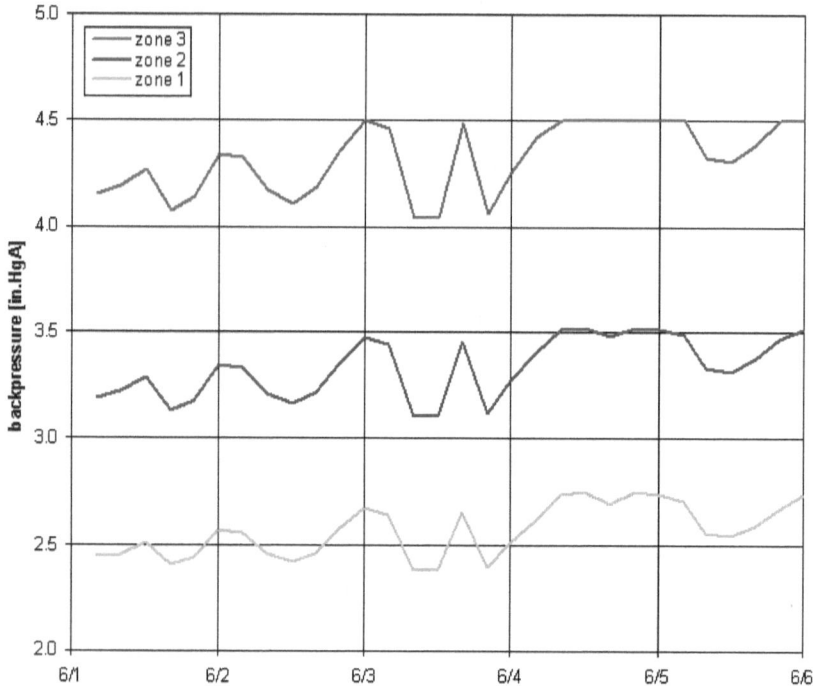

The cooling towers at Watts Bar can achieve a closer approach (i.e., cold water exiting temperature minus ambient wet-bulb temperature) than the spray cooling system used at the paper mill and proposed at Wolf Creek. This means that the cooling towers provide more hours of full load operation than spray cooling would. Of course, the cooling towers cost a whole lot more too plus there's the maintenance to consider and they could blow over or fall down, should a tornado come through or an earthquake occur.

Chapter 10. Cloud Cover Simulation

As we will see in the next chapter, solar rays are not so uniform as often portrayed, even in areas known for clear skies. In fact, the incident irradiation at any location varies considerably with time and also space. While blanket (gross scale) solar data is readily available, local data is not. Most often a single sensor is ideally positioned at a site. Most sensors are also quite inaccurate. A single active tracking instrument may cost as much as $25,000. It is no wonder that one or only a few are deployed. This paucity of data may give the impression of much greater available power than can be utilized by a field of solar collectors.

Thus we consider cloud cover simulation. This has two primary uses: 1) modeling actual performance and 2) revealing the often disappointing truth to decision makers when the project falls far short of projections, which are most often based on clear skies and plentiful solar rays. The following is a single frame of one such effort that I prepared for just this reason. The entire animation can be viewed by following the link.

http://dudleybenton.altervista.org/solar/clouds.gif

This animation was created from actual clouds (shown in the next figure) converted to gray-scale. The overall brightness was adjusted to match a four-hour period of incident solar data collected from an array of sensors. The resulting screen-like filter was then rolled diagonally across a photograph of the collectors taken at a cloudless moment. The direction of movement was in line with the prevailing wind at the site. All of the files and code (clouds.c) can be found in the online archive in folder examples\clouds.

The shading code is quite simple:

```
for(k=0;k<gif.frames;k++)
  {
  for(h=0;h<bm->biHeight;h++)
    {
    for(w=0;w<bm->biWidth;w++)
      {
      i=wide8*((h+4*k)%Shading->biHeight)
       +(w+4*k)%Shading->biWidth;
      j=shading[i];
      shade=(2500*((int)RedRGB(palette[j])))/3014;
      j=wide24*h+3*w;
      r=collectors[j];
      g=collectors[j+1];
      b=collectors[j+2];
      r=(((WORD)r)*((WORD)shade))/255;
      g=(((WORD)g)*((WORD)shade))/255;
      b=(((WORD)b)*((WORD)shade))/255;
      bits[j]=r;
      bits[j+1]=g;
      bits[j+2]=b;
      }
```

```
        }
    gif.bm[k]=BMPduplicate(bm);
    }
```

The incident solar irradiation (insolation) on each panel changes with time. If a single panel were connected to a dead (i.e., non-reactive) load, the voltage and current would vary with insolation; but this is not how such fields are operated. The collectors are connected in banks, which feed into multiple inverters (DC-to-AC electronic devices) and then into step-up transformers. Solar panels produce low-voltage direct current (DC); but the power grid requires high-voltage alternating current (AC). Transformers only operate on AC, not DC. The inversion process is necessary and the equipment is expensive. The panels are necessarily connected together at some level so that non-uniform insolation reduces the overall performance by more than simply the average, as would be measured by multiple sensors, which are not sharing the same load.

Chapter 11. Solar Irradiance Data

Solar power production is often quite disappointing due to rosy predictions that don't consider the realities of the sky. When evaluating the performance of solar energy collectors it is often assumed that the Direct Normal Irradiance (DNI) only varies with the position of the sun and is spatially invariant; but this is not the case, even in locations known for their clear skies.

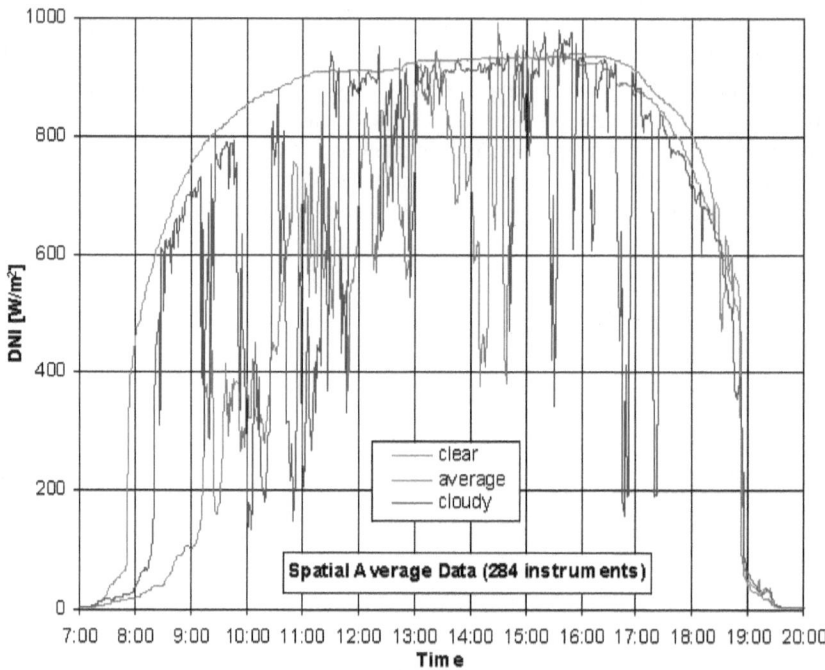

Three days have been selected from an extensive database: clear, average, and cloudy. As these three days are in the spring and almost contiguous, they illustrate how the DNI can vary considerably from day-to-day. There are 284 instruments recording one-minute data, for a total of 1,244,160 measurements. This first figure shows the average of all 284 instruments.

This second figure shows typical data from a single instrument. There is considerable variability, even on an average day.

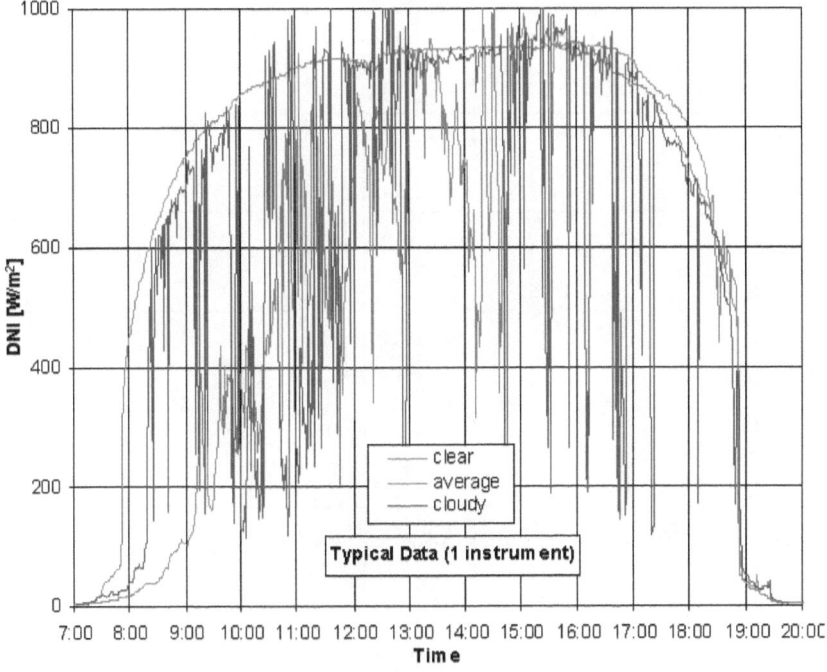

The third figure shows the standard deviation of DNI on the clear and average day. The standard deviation is a total mess on the cloudy day.

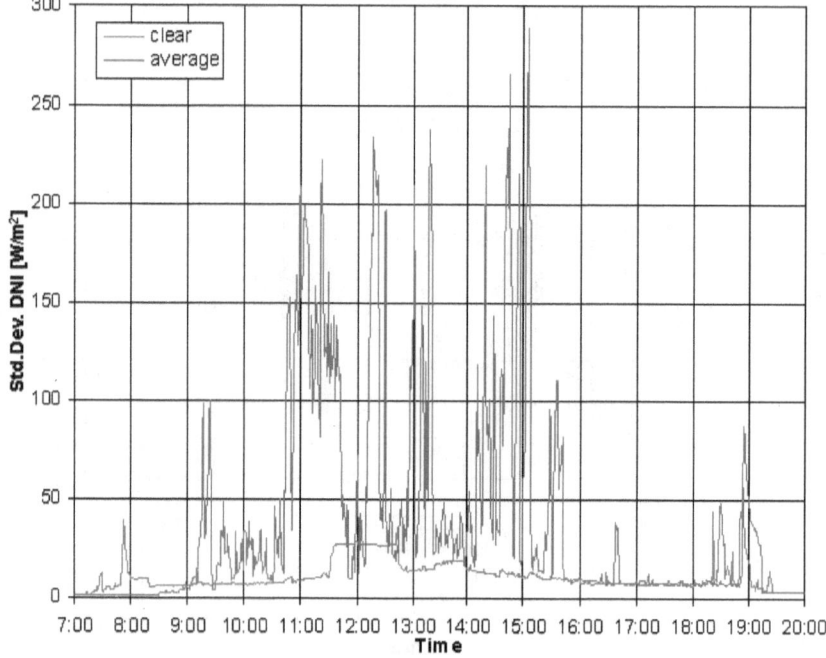

On this clear day the standard deviation approaches 7% of the average. These first figures only show the temporal variability.

The spatial variability of DNI at 12:52 on the clear day is 36 W/m² and is illustrated in the next figure:

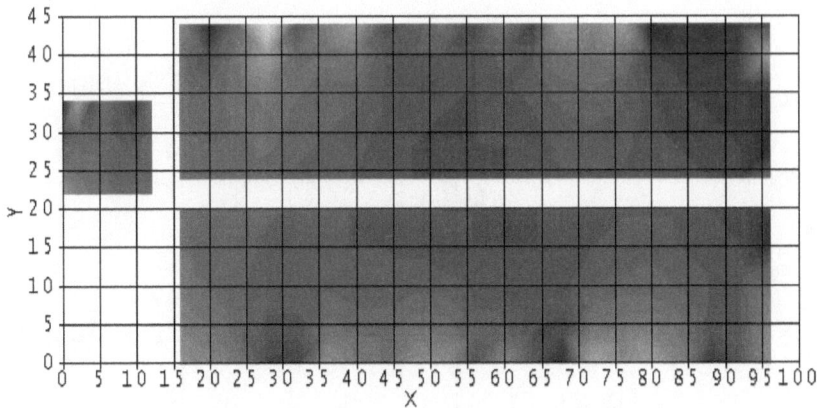

In the above figure the darkest shade corresponds to 909 W/m² and the lightest shade corresponds to 975 W/m². The standard deviation of DNI at 16:30 on the clear day is 12 W/m² and is illustrated in the next figure:

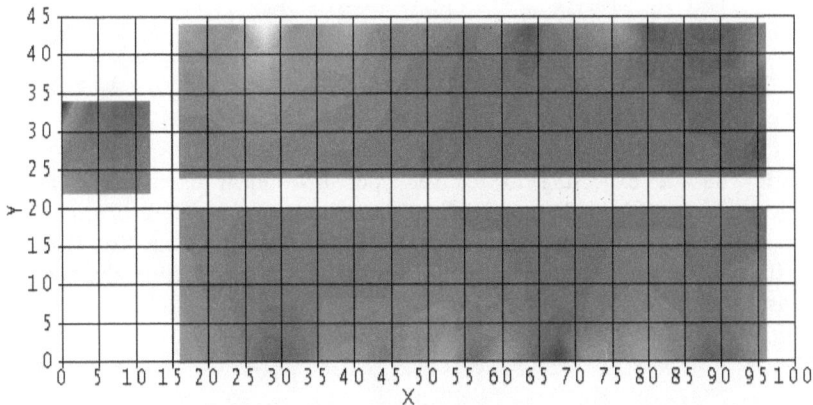

Three (very large) animations of this data with one frame per minute can be found at the following links:

http://dudleybenton.altervista.org/solar/clear.gif

http://dudleybenton.altervista.org/solar/cloudy.gif

http://dudleybenton.altervista.org/solar/average.gif

The expected power output of this 500-acre facility for these three days is shown in the next figure.

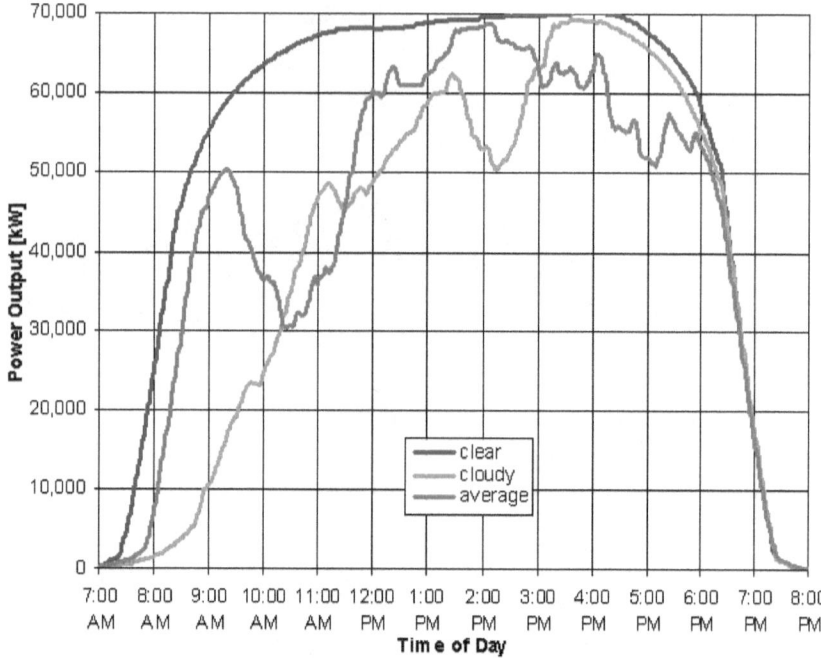

If you are expecting to recover your investment based on clear-day performance, you will disappointed, as the average is only 81% of this. On a somewhat cloudy day the output is only 74% of that on a clear day. Phoenix, Arizona is known for it's clear skies, but this is an anecdotal evaluation, not measured data, as in this next figure:

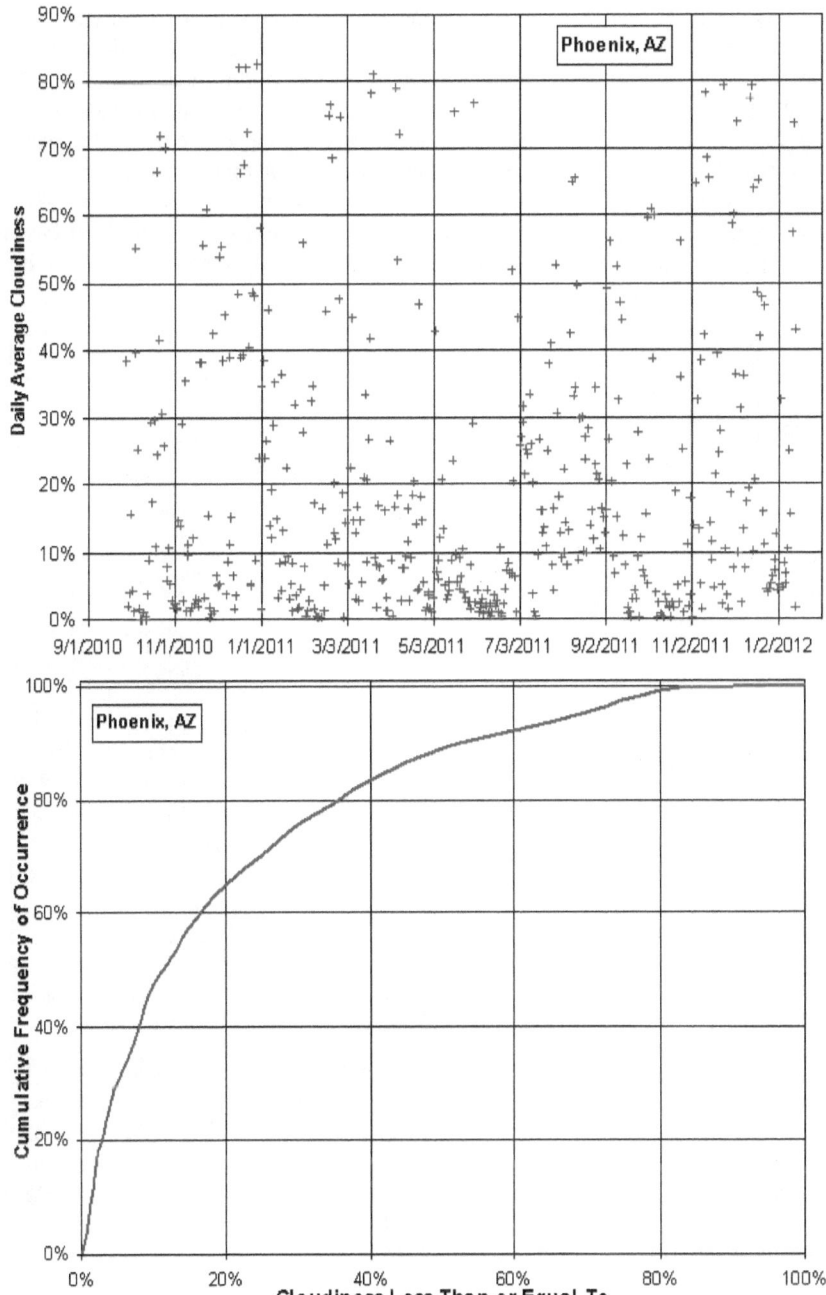

As these data show, it is naive to presume clear days, even in Phoenix. You can easily find equations for the position of the sun that account for time of day, day of year, and location on the surface of the Earth. You can also find equations for solar panel output as a function of DNI. What is often lacking is the fraction of the total incident radiation above the atmosphere that will make it to your solar panels. You will need historical data for this. In the next chapter we will consider how to construct solar data given available information. This next figure shows 233 days of 15-minute average data for DNI and GHI based on approximately 280 instruments reporting along with the weekly running average.

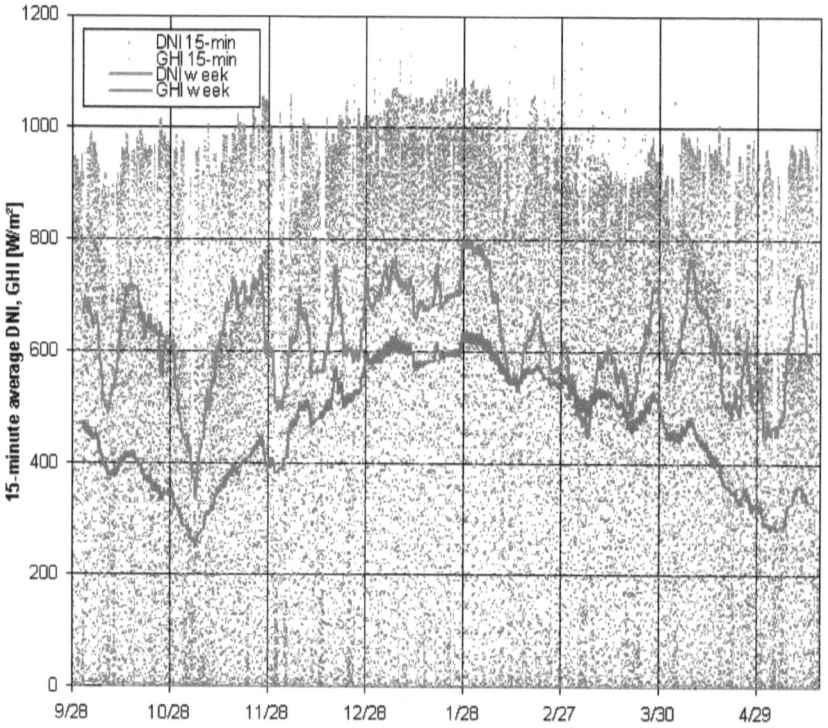

After curve-fitting two weeks of continuous data, the regression model for this plant was applied to the previous 8 months of data to arrive at a longer term performance factor. The clouds so impacted DNI that GHI, which was also recorded but with fewer instruments, was factored into the regression after noticing that the GHI sometimes exceeded the DNI (green dots above red ones) in the preceding figure. Approximately 40% of the difference was seen in the capacity during such times; hence the adjustment.

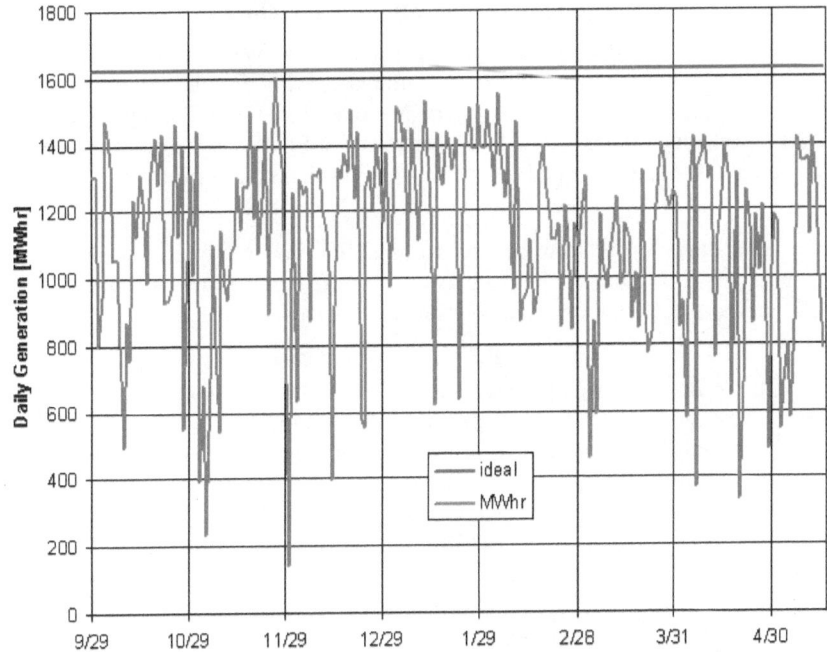

Over this time period, the collector field did approach the expected peak output (1625 MWhr/day), but not for more than a few hours so that the green curve never quite touches the blue line in the above plot. The average capacity over this period was estimated at 1126 MWhr/day or only 69.3% of the "design" value. If this hadn't been a solar plant, there would have been a huge lawsuit. No one who buys gas turbines would tolerate 69.3% of expected performance.

Chapter 12. Solar Plant Simulation

In order to simulate a more general solar power plant we will first need a formula for calculating the position of the Sun with time accounting for the location on the surface of the Earth. The National Renewable Energy Laboratory (NREL) operates under the U.S. Department of Energy (DoE) and is a good source for information related to solar energy. Their web site is:

https://www.nrel.gov/

The calculation is quite involved—assuming you need extreme precision. At least one document on this site is entirely devoted to the subject.[21] Source code for the calculation can also be found on this site.

https://midcdmz.nrel.gov/spa/

This source consists of nearly 1000 lines of code, not counting comments. While there is some infinitesimally small likelihood that you might need to know the position of the Sun to six significant figures, this is absurdly trivial compared to the haze and cloud obscuring factors shown in the previous chapter. Centro de Investigaciones Energéticas, Medioambientales y Tecnológicas (CIEMAT), has a much simpler and fully adequate calculation that is discussed in Appendix F.

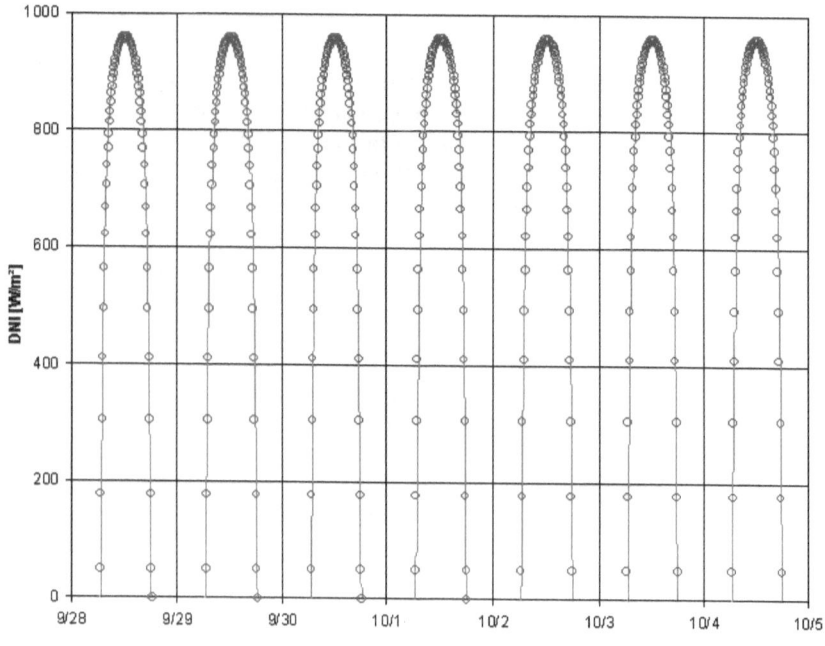

[21] Reda, I. and A. Afshin, "Solar Position Algorithm for Solar Radiation Applications," NREL/TP-560-34302, 2008.

In order to run a long-term simulation of a solar plant at some potential location, we need to know two things so as to characterize the incident power and ultimately the net power produced for an array of a particular size: clear sky DNI and cloudiness (including haze). The solar position can be calculated as discussed in Appendix F. The daily variation is shown in the graph on the previous page. The solid red curves are obtained through a long and tedious calculation. The blue dots are an empirical best-fit:

$$DNI = \frac{a}{e^{\left(\dfrac{b}{\cos\left(\dfrac{\pi(h-\tau)}{12}\right)}\right)}} \qquad (12.1)$$

where a=1180, b=0.2061, τ=12.25, and h is the hour (i.e., 0-23). This can be found in the solar_data.xls spreadsheet, along with the corresponding data. The timing of the peak (τ) varies throughout the year, as does the width of the peak (b) and the magnitude (a). The peak (clear sky) DNI varies throughout the year. For the selected site (Phoenix, AZ), the curve is:

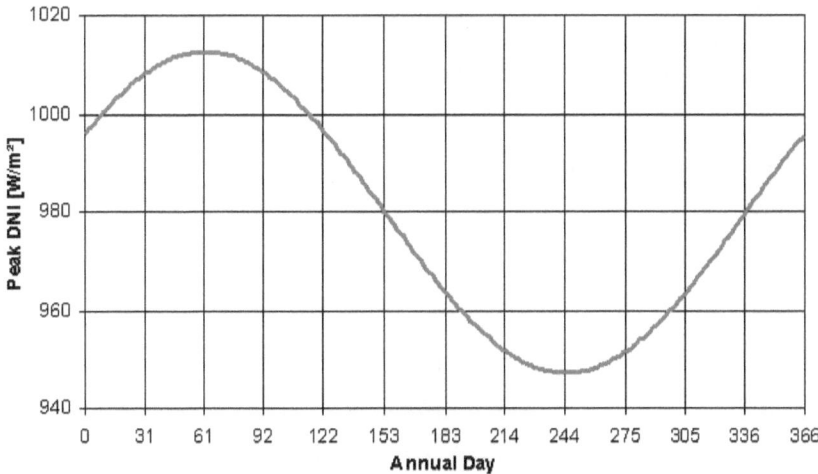

The solar "noon" (i.e., time of peak DNI) varies slightly throughout the year, as shown in this next figure.

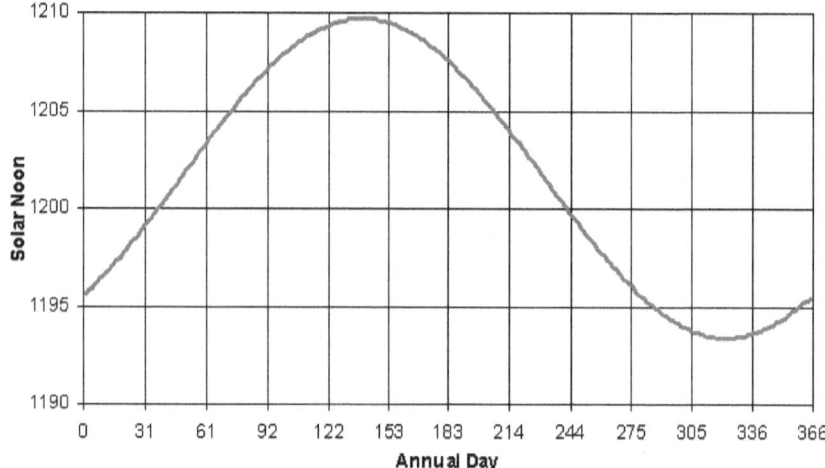

The sharp rise in DNI in the morning and fall in the afternoon also changes throughout the year. The spread in hours between the two curves in the figure below represents the daily solar power generating period.

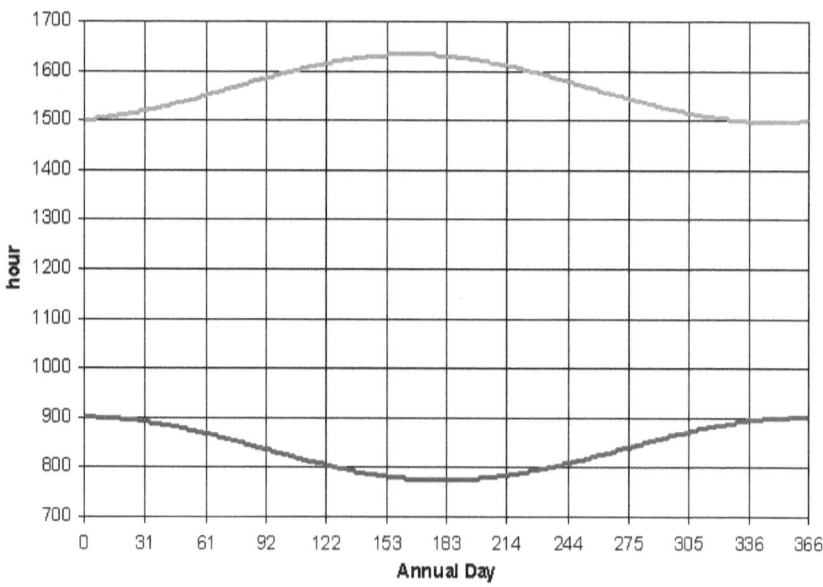

The following functions accurately represent the preceding curves and combine to yield a calculation of clear-sky DNI:

```
double DNIearly(int aday)
  {
  return(838.4+63.06*cos(2.*M_PI*(aday-363)/365.25));
  }

double DNIlate(int aday)
  {
  return(1565.2+68.14*cos(2.*M_PI*(aday-165)/365.25));
  }

double DNIpeak(int aday)
  {
  return(980.*(1.+0.033412*cos(2.*M_PI*(aday-
  64)/365.24)));
  }

double DNIclear(int aday,int hour,int minute)
  {
  double k=0.163,early,late,noon,peak,theta,time;
  early=DNIearly(aday);
  late=DNIlate(aday);
  noon=(early+late)/2.;
  peak=DNIpeak(aday);
  time=hour*60+minute;
  theta=M_PI*(time-noon)/(late-early);
  if(fabs(theta)<0.999*M_PI/2.)
    return(peak*exp(k)/exp(k/cos(theta)));
  return(0.);
  }
```

While there is considerable scatter, the average cloudiness does exhibit a trend over the year of available data:

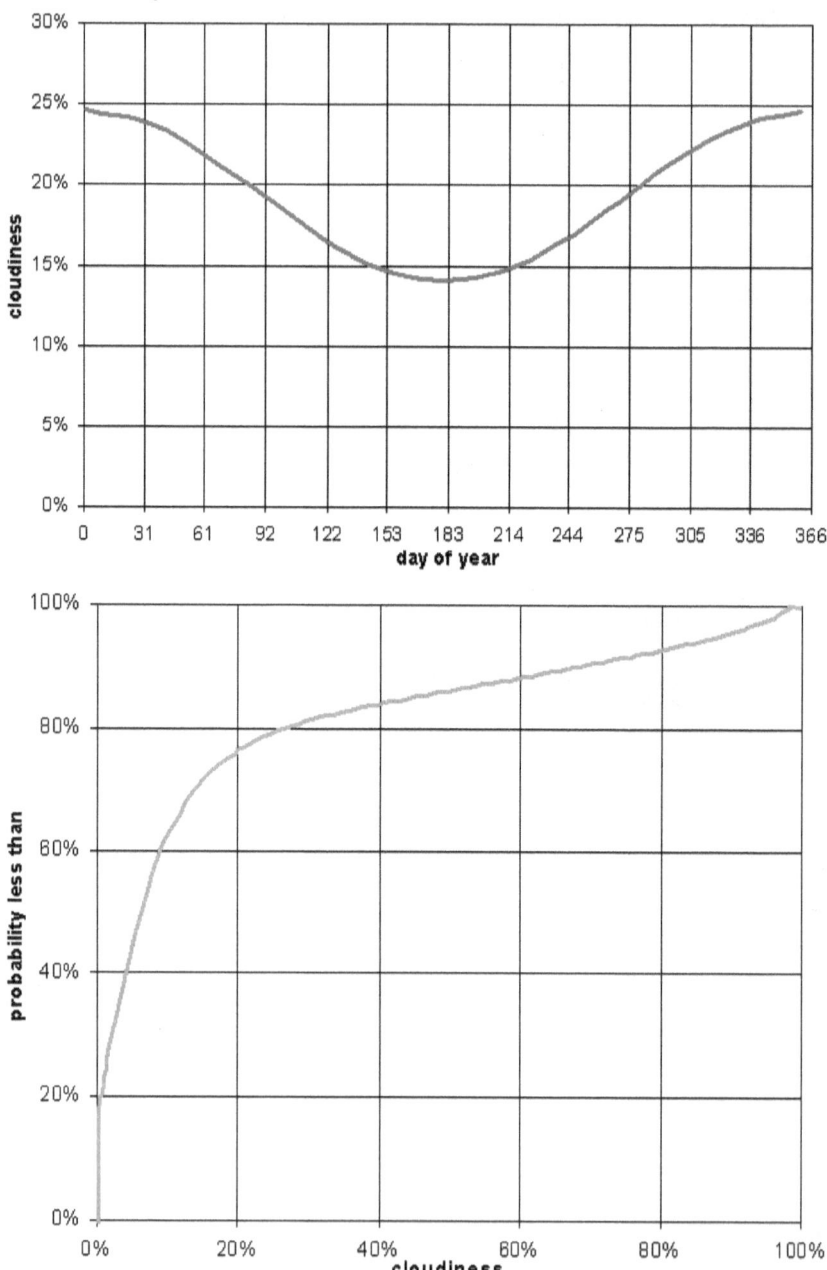

Clearly the cloudiness is not normally distributed (i.e., bell-shaped curve). The probability density is shown in this next figure:

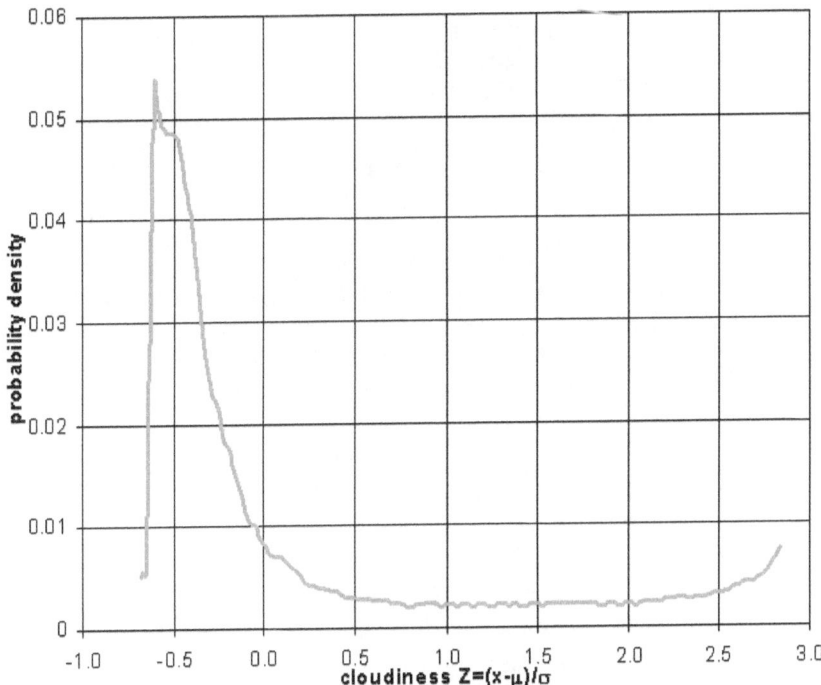

Here μ is the mean, σ is the standard deviation, and Z is the dimensionless factor often called a "score", as it is used for testing. There is too much scatter to infer any consistent daily trend in cloudiness but the probability of cloudiness is easily calculated from the data, as shown in the preceding figure. We can combine these curves plus Equation 12.1 and a random number generator to create a partial Monte Carlo simulation of the solar field. The clear-sky DNI will be deterministic, ultimately based on the latitude and longitude and the cloudiness will be random, but based on the probability curve.

Estimating Cloudiness

We use a uniformly-distributed random number to estimate cloudiness from a curve-fit of the cumulative probability less than. We do not use a normally-distributed random number, as this would presume a normal distribution, which we do not have in this case. Instead, the curve-fit links the random number to the appropriate cloudiness. The C <stdlib.h> function rand() returns a uniformly-distributed random integer between 0 and 32767; therefore, we use the following statement:

```
cloudiness=Cloudiness(rand()/32767.);
```

77

The cloudiness curve-fit is a rational polynomial (i.e., p(x)/q(x)):

```
double Cloudiness(double probability)
  {
  return(((((0.76288721101*probability
    -1.28320573162)*probability
    +0.624789144234)*probability
    -0.0558391005561)*probability
    -0.000509558186569)/((1.25994765396
    *probability-2.21186907023)
    *probability+1.)));
  }
```

Clouds do not instantly appear and disappear, so in order to achieve a reasonable approximation, we must incorporate a response time. This is easily done with a rolling average. A reasonable time constant based on historical data at this location is 3 hours. The simulation (less the DNI and cloudiness functions) is only a few dozen lines of code:

```
void Simulation(char*fname)
  {
  int day,hour,julian,minute,mo,month=0,y=0,year,yr;
  double area=75.,cloudiness=0.,DNI,MW,MWhr=0.,t=1./60./3.;
  FILE*fp;
  printf("Solar Collector Field Simulation\n");
  printf("output file: %s\n",fname);
  if((fp=fopen(fname,"wt"))==NULL)
    return;
  for(year=2019;year<2020;year++)
    {
    for(julian=JulianDay(year,1,1);
      julian<=JulianDay(year,12,31);julian++)
      {
      mo=month;
      yr=y;
      YearMonthDay(julian,&y,&month,&day);
      if(day==1&&mo>0)
        {
        printf("%i/%i,%.0lf\n",mo,yr,MWhr);
        MWhr=0.;
        }
      for(hour=0;hour<24;hour++)
        {
        for(minute=0;minute<60;minute+=15)
```

```
      {
      DNI=DNIclear(julian-JulianDay(year,1,1),
      hour,minute);
      cloudiness=(1.-t)*cloudiness
        +t*Cloudiness(rand()/32767.);
      MW=area*DNI*cloudiness;
      MWhr+=MW/60./1000.;
      fprintf(fp,"%i/%i/%i %i:%02i,%.01f\n",
        month,day,year,hour,minute,MW);
      }
    }
  }
}
printf("%i/%i,%.01f\n",mo,yr,MWhr);
fclose(fp);
```

} Julian is used to facilitate stepping through a sequence of days. A function is provided for this (in solar.c) and also the day that Microsoft® uses in Excel® (i.e., the number of days since 1/1/1900). Output for one year by month (total MWhr/month) is:

```
Solar Collector Field Simulation
1/2019,219
2/2019,236
3/2019,287
4/2019,290
5/2019,320
6/2019,328
7/2019,319
8/2019,299
9/2019,261
10/2019,257
11/2019,231
12/2019,225
```

The first 3 days of 15-mimute simulation looks very much like the measured data and ironically contains a cloudy, average, and clear day:

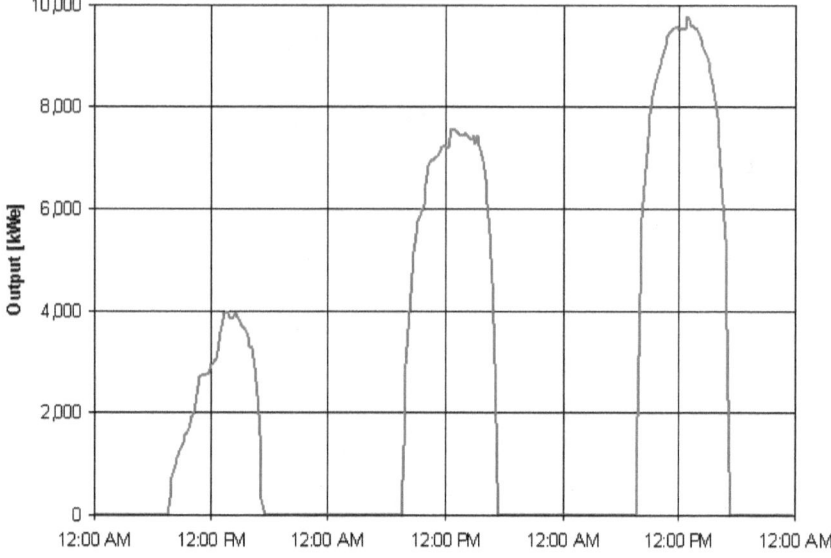

Appendix A. Select Publications

The following publications are available on the Web. All TVA reports are available through the Freedom of Information Act (FOIA), as these are US government documents. These are divided into two categories: 1) historical and projected response and 2) predicted response to potential changes in climate.

"A Supplemental 316(a) Demonstration for Alternative Thermal Discharge Limits for Browns Ferry Nuclear Plant, TVA Report ONP-83471, 1983, with P. Ostrowski et al. This study quantifies the environmental impact of plant operations within the context of the National Pollutant Discharge Elimination System (NPDES) provisions within the Clean Water Act of 1977, focusing on Section 316(a), which makes provision for exceptions to the uniform restrictions. Historical meteorological data were used throughout. http://dudleybenton.altervista.org/pub/316A.pdf

"Analysis of Browns Ferry Nuclear Plant Operation with Four and Six Cooling Towers," TVA Report WR28-2-67-122, 1986, with W. G. Carpenter. This study was initiated in response to a debilitating fire that occurred at the plant, completely destroying one and partially damaging a second cooling tower. Historical meteorological data were used throughout the simulations. http://dudleybenton.altervista.org/pub/67-122.pdf

"Analysis of Shell and Tube Heat Exchangers for Use in OTEC Systems," 15th Southeastern Seminar on Thermal Sciences, 1979. This study was undertaken to quantify the power generation potential for utilizing the thermal gradients available in the world's oceans and the Gulf Stream in particular. http://dudleybenton.altervista.org/pub/OTEC.pdf

"Analysis of the Paradise Fossil Plant Cooling System: The Impact on Generation of Additional Cooling Towers," TVA Report WR28-2-64-122, 1993. This study provided a basis for making an economic decision whether to erect additional cooling towers at the plant. Historical meteorological data were used. http://dudleybenton.altervista.org/pub/64-122.pdf

"Cursory Analysis of the Condition of the Cooling Towers at the Paradise Steam Plant," TVA Report No. WR-28-2-64-104, 1986. This study assessed the condition and the impact on operations using historical meteorological data. http://dudleybenton.altervista.org/pub/64-104.pdf

"Impact of Atmospheric Lapse Rate on Power Plant Performance," American Power Conference, 1992. This paper describes how one particular type of atmospheric phenomenon impacts the performance of a natural draft cooling tower. http://dudleybenton.altervista.org/pub/LAPSE.pdf

"Impact of Incremental Changes in Meteorology on Thermal Compliance and Power System Operations," TVA Report WR28-1-680-109, 1992, with B. Miller et al. This study quantified the performance of eleven different power

81

systems to changes in meteorology and climate-related regulations. http://dudleybenton.altervista.org/pub/680-109.pdf

"Impacts of Changes in Air and Water Temperature of Thermal Power System Generation," AWRA, 1992, with B. Miller et al. This is a summary of the more extensive previous work (WR28-1-680-109), focusing primarily on climate change. http://dudleybenton.altervista.org/pub/TPGEN.pdf

"Integrated Assessment of Temperature Change Impacts on the TVA Reservoir and Power Supply Systems," Proceedings of the Water Forum: Saving a Threatened Resource—In Search of Solutions, American Society of Civil Engineers, 1992, with B. A. Miller, V. Alavian, M. D. Bender, P. Ostrowski, Jr., J. A. Parsly, and M. C. Shiao. This paper came out of a large team effort focused on alternative strategies for resource utilization. http://dudleybenton.altervista.org/pub/RPSS.pdf

"Modeling the Response of a Multi-Unit Electric Power Plant to a Changing Environment," Tennessee Water Resources Symposium, 1992. This paper describes the software I developed to simulate multiple different power stations. http://dudleybenton.altervista.org/pub/MUPIT.pdf

"Paradise Steam Plant Cooling Tower Operation Scheduling Program," TVA Report WR28-2-64-102, 1983. This is a description and validation of the software developed to continuously schedule operation of the plant. http://dudleybenton.altervista.org/pub/64-102.pdf

"Predicted Effects for Mixed Temperatures Exceeding 30C (86F) in Guntersville Reservoir TVA Report ONRED-8286, 1982, with D. A. McIntosh et al. This study combined several simulation models in order to quantify the impact of a proposed regulation on Browns Ferry Nuclear Plant operations. http://dudleybenton.altervista.org/pub/GUNT.pdf

"Predictions of Load Reductions for Browns Ferry Nuclear Power Plant to Comply with the Alabama Thermal Water Quality Standards," TVA Report WR28-2-64-136, 1980, with W. L. Harper. This study was undertaken to quantify the impacts of proposed changes to environmental regulations. http://dudleybenton.altervista.org/pub/67-106.pdf

"Results of Analysis of the Bellefonte Nuclear Plant Heat Rejection System," TVA Report WR28-2-88-114, 1992. This study evaluated the expected shortfall in performance of the cooling tower and how this would impact operations. http://dudleybenton.altervista.org/pub/88-114.pdf

"Results of Analysis of the Watts Bar Nuclear Plant Heat Rejection System," TVA Report WR28-2-85-136, 1992. This study was undertaken to quantify the impact of and predict the revised performance of the plant based on design deficiencies in major components provided by one particular manufacturer. http://dudleybenton.altervista.org/pub/85-136.pdf

"Sensitivity of the TVA Reservoir and Power Supply Systems to Changes in Meteorology," National Conference on Climate Change, 1992, with B. Miller et al. This study was based on the same eleven systems, fine tuned to focus on what was considered to be the most likely scenarios for climate change at the time. http://dudleybenton.altervista.org/pub/TVAPS.pdf

"Sensitivity of the TVA Reservoir and Power Supply Systems to Extreme Meteorology," TVA Report No. WR28-1-680-111, with B. A. Miller et al. This study was preliminary to and provided a basis for the later studies listed previously that dealt specifically with climate change. http://dudleybenton.altervista.org/pub/680-111.pdf

Appendix B. Global Surface Summary of the Day

Global Surface Summary of the Day is derived from The Integrated Surface Hourly (ISH) dataset. The ISH dataset includes global data obtained from the United States Air Force (USAF) Climatology Center, located in the Federal Climate Complex with National Climate Data Center (NCDC) in Asheville, North Carolina. THE NCDC is operated by the National Oceanic and Atmospheric Administration (NOAA), which is under the U.S. Department of Commerce (DoC). The data are located on a server and accessible to all. The general website link is:

https://catalog.data.gov/dataset/global-surface-summary-of-the-day-gsod

Data can be found on the ftp site:

ftp://ftp.ncdc.noaa.gov/pub/data/gsod/

Stations are located around the globe. The current coverage map is:

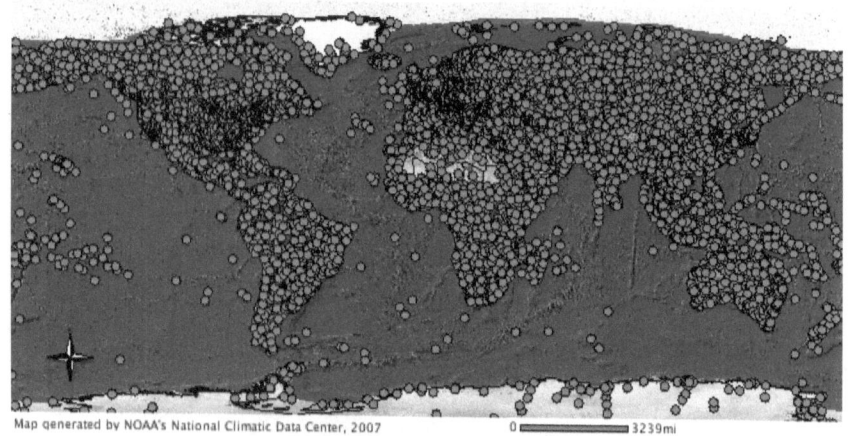

Map generated by NOAA's National Climatic Data Center, 2007   0 ▬▬▬▬ 3239mi

Documentation and a list of stations can be found in the root folder, including files: country-list.txt, GSOD_DESC.txt, GSOD-IMPROVEMENTS.TXT, ish-history.csv, ish-history.txt, readme.txt.

Subfolders are arranged by year. Within the subfolders are files for each station by number and also collections (LINUX tar files) for the entire (or partial) year, including all stations (1901.tar through 2019.tar). These must be "un-tarred" (you can use the utility by that name provided in my online archive) and then "un-gzipped" (you can use the gzip utility provided in my online archive). The NCDC does not provide utilities for the Windows® operating system and I don't use UNIX (or LINUX). The NCDC was operational long before Windows® Server was released or any Windows-based platform could possibly have handled the sheer quantity of data; thus, it is UNIX-based.

As I have used this resource for many years, I have a number of utilities to facilitate preparation of the data. While some stations in this repository have

84

data as far back as 1901, these are few and far between. If you are interested in geographical and temporal extent of data coverage, consider the following:

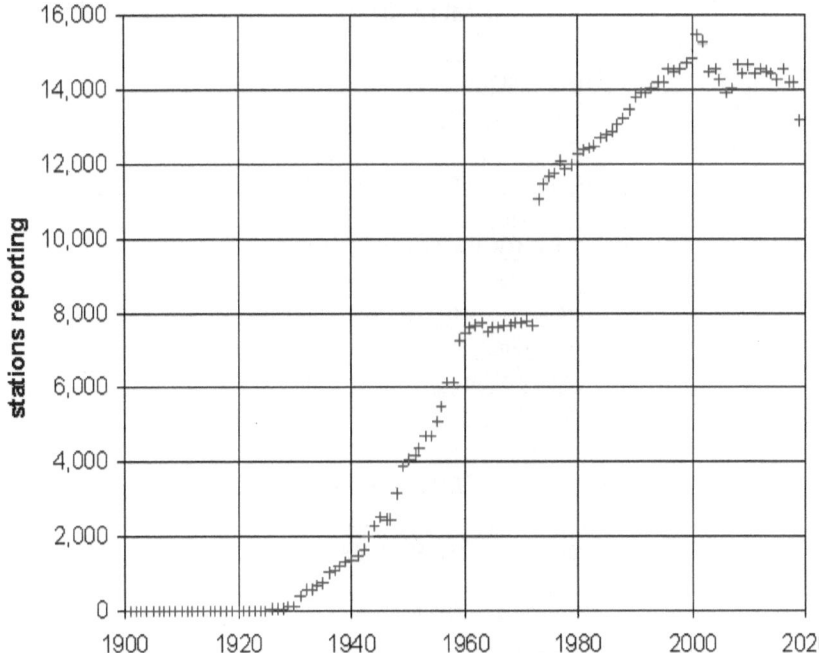

The number of stations reporting greatly increased in 1973 and leveled-off in 2000. I have used this data since 1980 to perform global and local analyses. Several tools are provided in the online archive in folder examples\GSOD. One of these is a spreadsheet (isd-history.xls) that can be easily updated with the latest file from NCDC (isd-history.csv) and be used to find the stations closest to any particular location on Earth. For example, a new power plant is under construction in in Dar es Salaam, Tanzania. You might need historical meteorological data appropriate for the Songas Ubungo Plant, located at 6.7924° S, 39.2083° E. The six closest stations are:

USAF	WBAN	LAT	LON	ELEV(M)	BEGIN	END	distance
638940	99999	-6.878	39.203	55.5	19500201	20190708	0.1
638700	99999	-6.222	39.225	16.5	19570101	20190708	0.6
638950	99999	-7.917	39.667	21.0	19570101	20141221	1.2
638660	99999	-6.833	37.650	526.0	19570101	20190708	1.6
638450	99999	-5.257	39.811	24.4	19570101	20190424	1.6
638440	99999	-5.092	39.071	39.3	19570101	20190124	1.7

The utility (getstn.c) extracts the available data from 638940-99999, which is a CSV (comma separated values) file of approximately 2MB in size. The data consists of approximately 17,500 days from 1/1/1957 through the present. The data are arranged in columns as listed in the following table:

Table A1. GSOD Header	
column	description
USAF	USAF station number
WBAN	WBAN station number
DATE	m/d/yyyy
TEMP	daily average dry-bulb temperature [°F]
NTEMP	number of readings
DEWP	daily average dew-point temperature [°F]
NDEWP	number of readings
SLP	daily average barometric pressure adjusted to sea level [mbar]
NSLP	number of readings
STP	daily average unadjusted barometric pressure [mbar]
NSTP	number of readings
VISIB	visibility [miles]
WDSP	daily average wind speed [knots]
NWDSP	number of readings
MXSPD	daily maximum wind speed [knots]
GUST	wind gust [knots]
TMAX	daily maximum dry-bulb temperature [°F]
TMIN	daily minimum dry-bulb temperature [°F]
PRCP	precipitation [inches]
SNDP	snow depth [inches]
FRSHTT	flags (fog, rain, drizzle, snow, ice, hail, etc.)

The first three entries are:

```
USAF,WBAN,YEARMODA,TEMP,NTMP,DEWP,NDEWP,SLP,NSLP,STP,NST
P,VISIB,WDSP,NWDSP,MXSPD,GUST,MAX,MIN,PRCP,SNDP,FRSHTT
638940,99999,01/01/1957,82.2,23,74.7,23,1009.8,23,9999.9
,0,23.1,23,4.2,23,9.9,999.9,90.0,75.9,0.00,999.9,000000
638940,99999,01/02/1957,80.9,24,74.5,24,1008.5,24,9999.9
,0,23.0,24,3.4,24,9.9,999.9,90.0,73.9,0.00,999.9,000000
638940,99999,01/03/1957,81.2,24,74.9,24,1008.6,24,9999.9
,0,23.9,24,4.2,24,14.0,999.9,90.0,73.9,0.00,999.9,000000
```

A selection of data for 2019 is shown in the following figure:

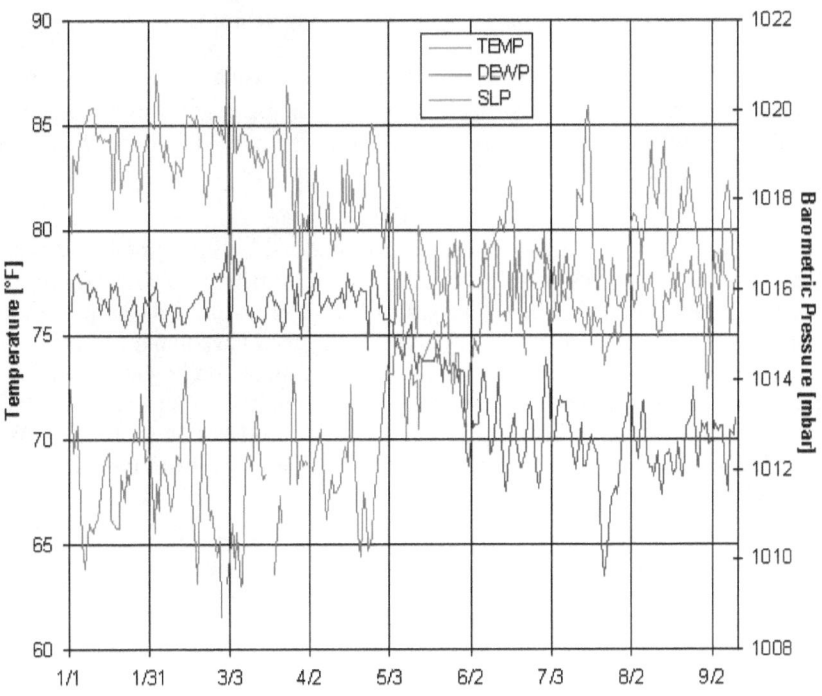

Appendix C: Disaggregating Daily Data

In Appendix B we extracted daily average meteorological data from the GSOD. This included a daily minimum and maximum dry-bulb temperature. Most often, hourly data are required to drive a simulation. An excellent reference on this topic is that of Waichler and Wigmosta.[22] This reference is readily available online and discusses several of the most important challenges as well as drawbacks of various approaches.

Temperatures

I have many sequences of hourly data collected by the National Weather Service (NWS) and also the Tennessee Valley Authority (TVA). There are various methods of estimating hourly data from daily. Although the process is not exact, I have selected from those methods the one I have found to work best. Consider two hours: one corresponding to the maximum daily temperature and the other corresponding to the minimum. These hours vary from day-to-day, but over a long period of time average approximately 2:00 PM and 2:00 AM. We interpolate over the day, requiring an exact match at these two points and also match the daily average.

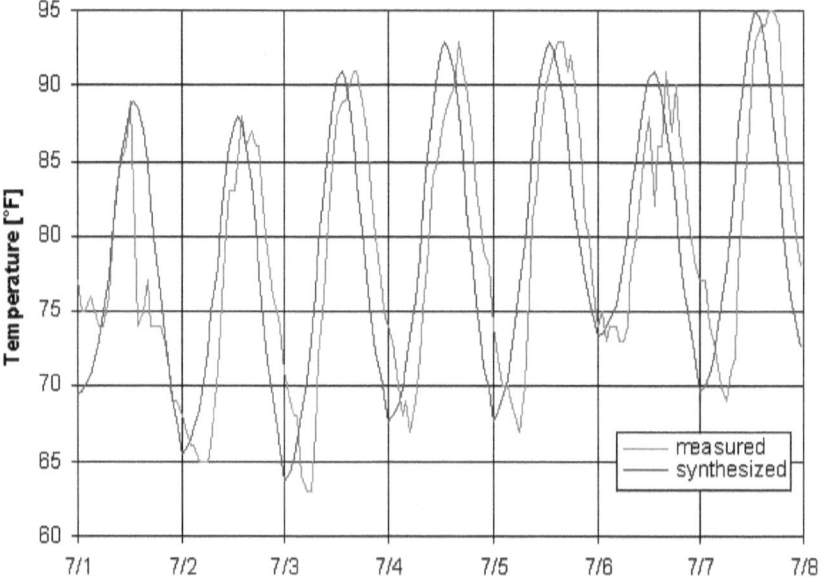

[22] Waichler, S. R. and M. S. Wigmosta, "Development of Hourly Meteorological Values from Daily Data and Significance to Hydrological Modeling at H. J. Andrews Experimental Forest," *Journal of Hydrometeorology*, Vol. 4, pp. 251-263, April 2003.
https://andrewsforest.oregonstate.edu/sites/default/files/lter/pubs/pdf/pub3636.pdf

We could be accomplished by a quadratic (i.e., T=a+b*t+c*t², where T is temperature, t is time, and a, b, and c are constants). This produce a smooth curve over a single day, but not from day-to-day. An exponential decay (i.e., a*exp(-t²/τ²)) away from each of the control points does assure a smooth transition. While this approximation is not perfect, it does agree fairly well with actual hourly data and has the advantage of never producing wild results— something that other interpolation schemes often do. This process is performed by the utility tohourly.c, which can also be found in the archive. Running tohourly on the output of getstn for this particular station expands the 17,404 days of data to 416,352 hours. The preceding figure shows one week of values calculated in this way. The equation is:

$$T = (1-x)\left[T_1 + \frac{h}{24}(T_2 - T_1)\right] + xT_X$$

$$x = e^{\left(\frac{h-13}{6.7}\right)^2}$$

(C.1)

where T_1 and T_2 are the minimum values for the current and following day, T_X is the maximum for the current day, and h is the hour (i.e., 0-23).

The dew-point and barometric pressure vary much more slowly than the dry-bulb. The dew-point is a direct indication of moisture content, which explains the slower response rate. Changes in barometric pressure produce and result from wind, which explains its slower response rate. While dry-bulb temperature is also impacted by wind, solar heating and nighttime heat loss are primarily responsible for the rate of change, which explains its faster response rate. The dry-bulb also tends to vary more rapidly during the day than night and the exponential decay approximation replicates this behavior.

Wind

Occasionally, you will need hourly wind speed, for example, to drive a cooling pond model. Typically, the least wind is at dawn and dusk. The simplest functional relationship that approximates this behavior is:

$$V = V_{MAX} \sin\left(\frac{2\pi(hour - 6)}{24}\right)^2$$

(C.2)

The average over a 24-hour period is 0.5. The GSOD contains the average and maximum wind speed; so you could use 2sin² times the average or sin² times the maximum to arrive at a meaningful disaggregation. You could also vary the hour of sunrise and sunset based on the day and latitude, even compensating for time zone differences and daylight savings, although weather data most often does not consider daylight savings time adjustments. The magnitude is illustrated in the following figure:

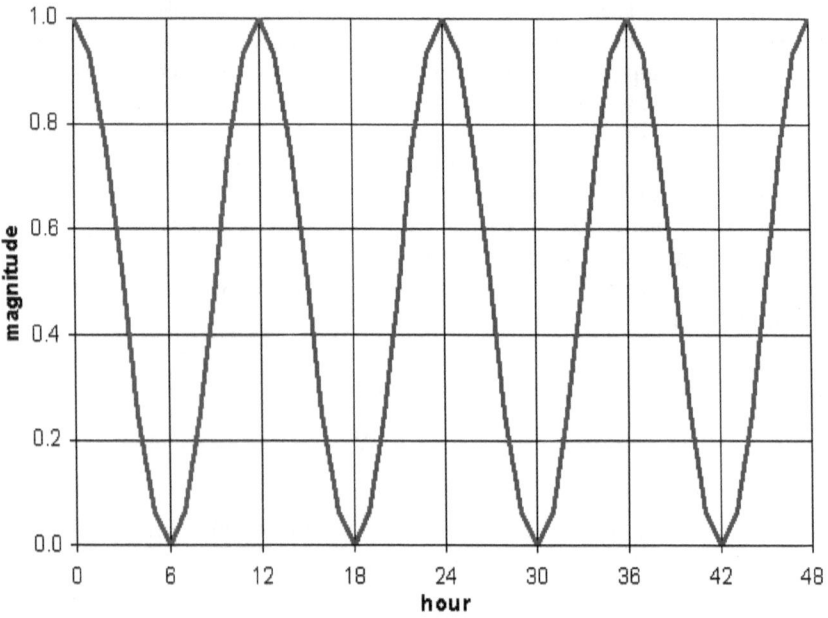

Appendix D. Interpolating Spatial Data

The first figure in Appendix B shows the distribution of meteorological stations around the globe. There are approximately 14,000 stations currently reporting data to the NCDC for the GSOD. This may be fully adequate for your needs; however, there are a few locations where the data is sparse or does not span enough years to be statistically significant. In that case, spatial interpolation may also be necessary. The following figure is more revealing than the one on the NCDC web site:

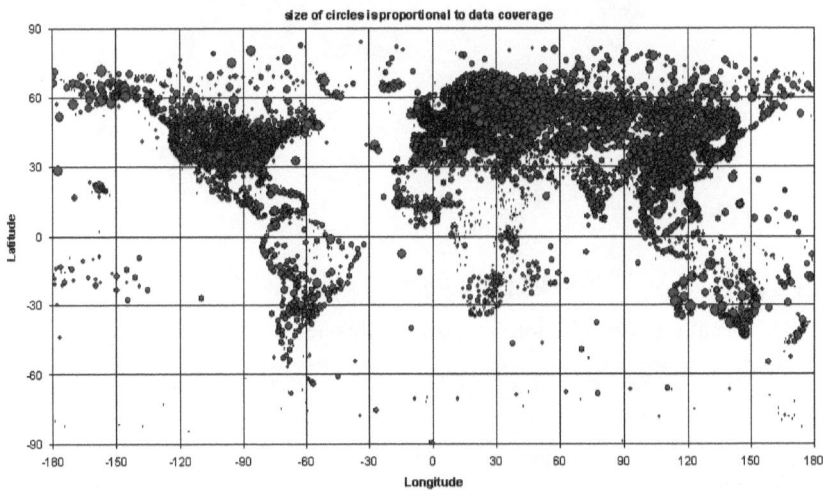

There are large areas in Africa and South America without a significant meteorological record—at least not in the GSOD. I created this figure by counting days of valid data in the entire 56GB available on the NCDC ftp site. While it might seem a daunting task to spatially interpolate each day and then temporally interpolate each hour in order to synthesize an adequate data set for a location within the interior of Brazil, Chad, Congo, Libya, Mali, Niger, Sudan, or Zambia, it's easier than you might think. I've done this many times and you're welcome to have the source code.

A variety of methods can be used to interpolate spatially, including: linear (which requires a grid), inverse distance, and kriging. I have found that inverse distance works best for this type of data. A typical global temperature profile (for 5/8/2018) is shown in this next figure:

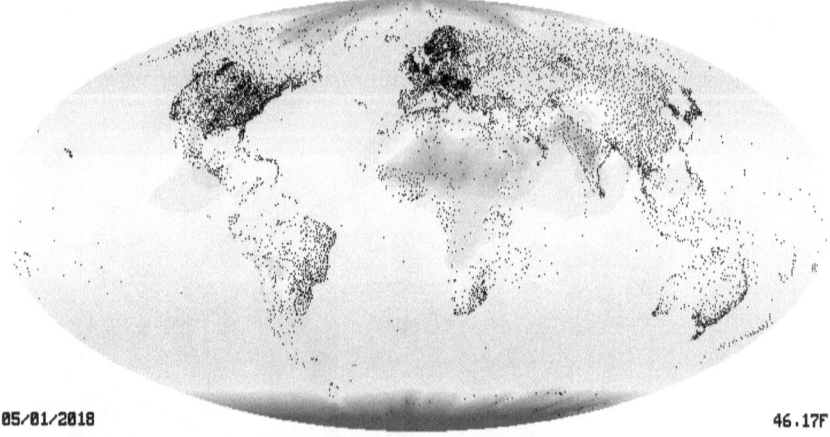

05/01/2018   46.17F

A typical grid (tessellation) is shown in this next figure:

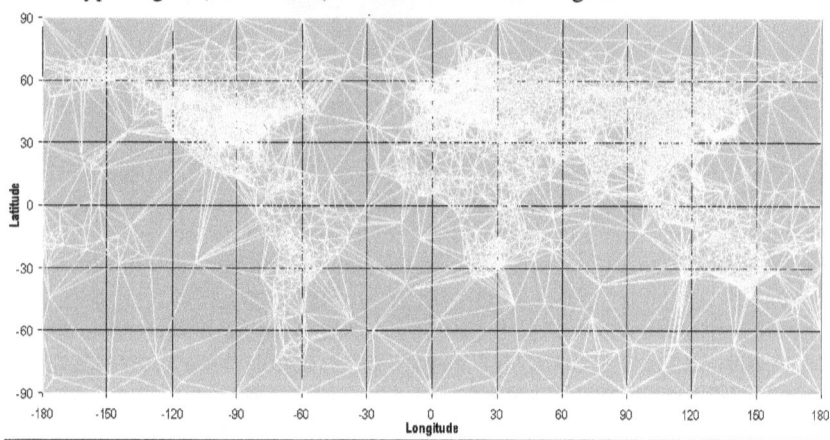

The same can be done with barometric pressure:

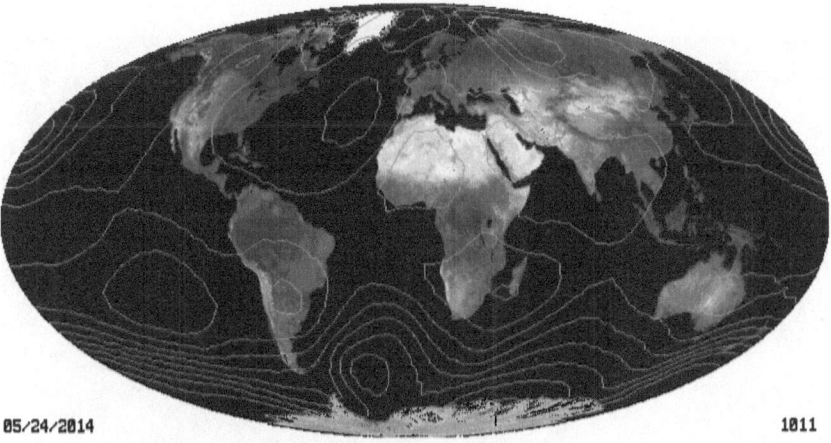

05/24/2014   1011

Rainfall data is often missing from the GSOD, but the same methodology can be applied. This particular day corresponds to a large tropical depression.

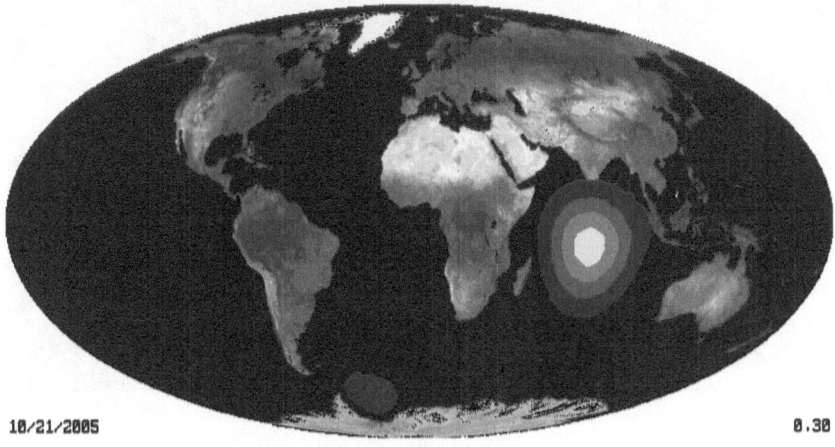

10/21/2005   0.30

The GSOD also contains wind data, which can be used to calculate vectors over the surface of the Earth:

07/02/2018 8.5

As indicated by these figures, various projections can be used to present the data, including: Mercator[23] and Mollweide.[24]

---

[23] A cylindrical map projection developed by Flemish geographer and cartographer Gerardus Mercator (1512-1594). This projection greatly exaggerates the size of objects near the poles, for instance, Greenland.

[24] An equal-area, pseudo-cylindrical projection often used for global maps of the world or night sky. Named after German mathematician and astronomer Karl Brandan Mollweide (1774–1825).

Appendix E. Natural Draft Cooling Towers

Predicting and presenting the performance of natural-draft cooling towers is considerably more complicated than mechanical-draft because there is an additional variable: ambient relative humidity. As a result of the Merkel approximation, the performance of mechanical-draft cooling towers can be accurately represented as depending on flow, range, and ambient wet-bulb. This is not true for natural-draft towers, as the heat and mass transfer impact the density, which drives the flow of air through the tower. The impact of ambient relative humidity may even be greater than that of range or flow in some circumstances.

The classical representation of natural-draft cooling tower performance is illustrated in the next two figures. The first shows the cold-water temperature vs. wet-bulb temperature for several values of ambient relative humidity. The second is the *range correction*, that is, the adjustment in cold-water temperature for range, also vs. ambient wet-bulb temperature.

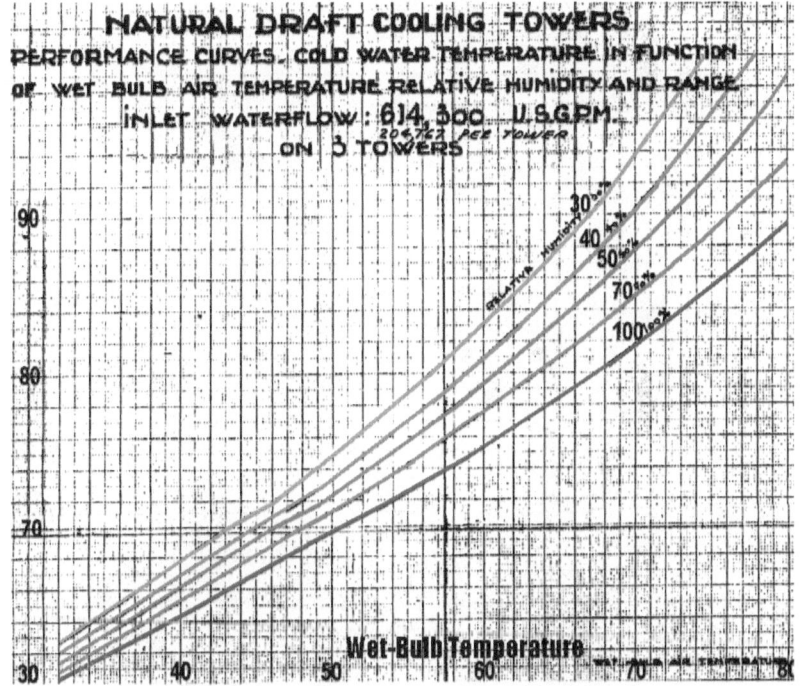

These curves are quite old (from the late 1960s), and are included here because such information is generally proprietary in nature. These curves, however, are for a U.S. government project (TVA's Paradise Steam Plant) and can be obtained through the Freedom of Information Act (FOIA).

95

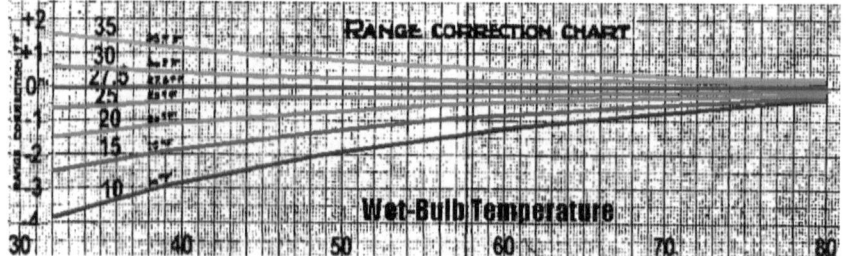

For this tower there are five sets of curves at five different water flows. These curves have been digitized and the range correction applied to each curve, resulting in over 15,000 points for cold-water temperature as a function of flow, range, ambient relative humidity, and ambient wet-bulb temperature. All of this data is included as an Excel® spreadsheet in the on-line archive for my book *Evaporative Cooling*. More information and discussion can be found therein. This relationship can be accurately approximated ($R^2=0.9969$) by a second-order multi-variable regression, as shown in this next figure:

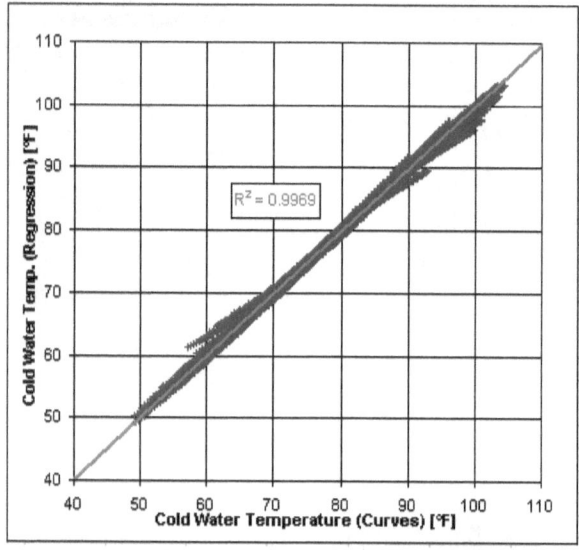

While such a regression is fully adequate for approximating the performance and correcting test data, it doesn't provide a meaningful visual presentation. A more creative presentation results from plotting curves on three sides of a cube, as illustrated in the next figure. Such a plot is usually drawn on a flat sheet of paper and then folded as indicated.

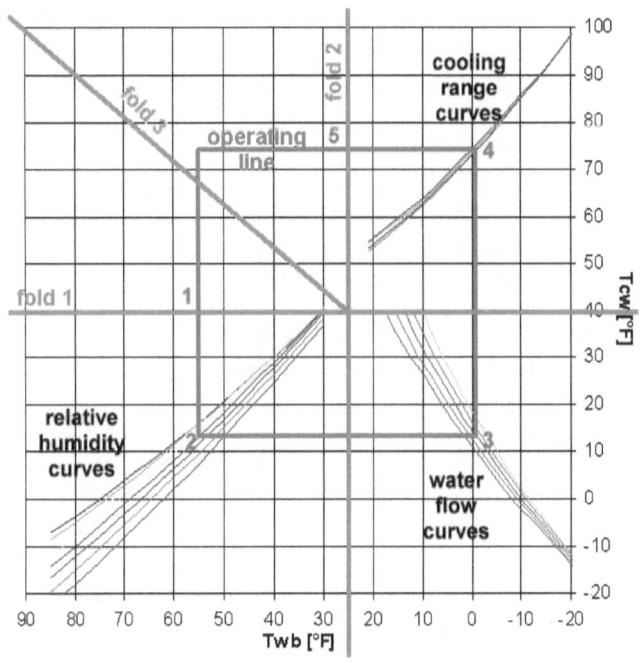

By folding along the lines as indicated above:

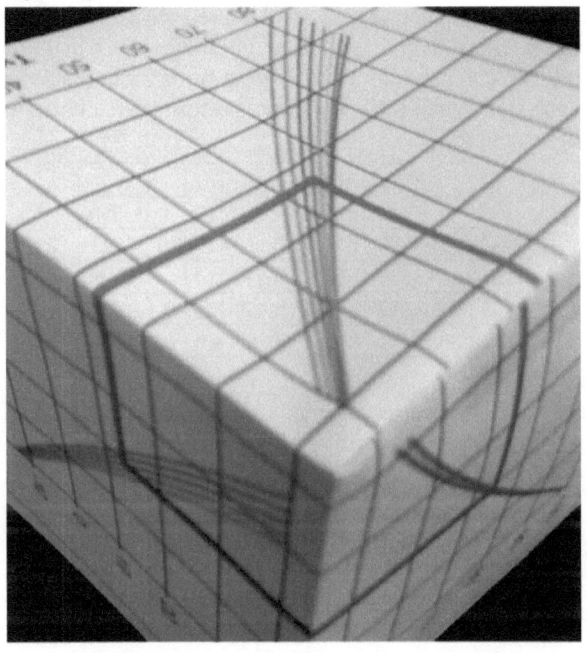

These curves representing the cooling tower performance are obtained by introducing two additional coordinates. The first will be given the symbol, X, and is the distance between points 1 and 2 on the previous graph. The second will be given the symbol, Y, and is the distance between points 2 and 3. These coordinates do not correspond to thermodynamic variables; they merely facilitate creation of the graph.

A typical operating point is found beginning at point 1 (the ambient wet-bulb) and drawing a vertical line down to point 2, where this intersects the ambient relative humidity curve. From point 2, a line is drawn horizontally to point 3, where it intersects the flow curve. From point 3, a line is drawn vertically to point 4, where it intersects the range curve. The cold-water temperature is then read from the scale at the right.

Each set of curves can be expressed in terms of a simple second order expansion. There are three sets of curves (relative humidity, flow, and range) and seven variables (wet-bulb, relative humidity, X, flow, Y, range, and cold-water temperature). Their inter-relation is given by the following equations:

$$X = a_1 + a_2 RH + a_3 Twb + a_4 RH^2 + a_5 RH \cdot Twb + a_6 Twb^2 \qquad \text{(E.1)}$$

$$Y = b_1 + b_2 flow + b_3 X + b_4 flow^2 + b_5 flow \cdot X + b_6 X^2 \qquad \text{(E.2)}$$

$$Tcw = c_1 + c_2 range + c_3 Y + c_4 range^2 + c_5 range \cdot Y + c_6 Y^2 \qquad \text{(E.3)}$$

There are eighteen constants in these three equations, but these are not all free. In order for the temperature scale of the three sets of curves to be the same, the coefficient b_3 must equal one. The coefficients a_1 and c_1 control the horizontal and vertical gaps, or the distance from each set of curves to the corner. A regression performed on these performance curves produces the following coefficients:

a1	70	b1	-1.0061	c1	70
a2	13.77244	b2	-11.3479	c2	3.433356
a3	-1.38779	b3	1	c3	-1.26823
a4	-14.8557	b4	1.756232	c4	-0.2345
a5	0.316842	b5	-0.24559	c5	0.131833
a6	0.002149	b6	-0.00385	c6	0.006758

This provides a convenient representation of an existing tower design, but doesn't explain how the tower is designed in the first place. This particular tower was designed by Marcel LeFevre using Merkel's method.

Appendix F. Calculating Solar Position

While the NREL solar position formulation of Reda and Andreas might achieve a very high level of precision, this is completely unnecessary for solar calculations, as haze and cloud cover have many orders of magnitude greater impact on any calculations. CIEMAT provides a much simpler and fully adequate calculation. The C^{++} source code can be downloaded from:

http://www.psa.es/sdg/archive/SunPos.cpp

This has been modified for use as an Excel® macro that can be found in the online archive in the folder examples\solar in spreadsheet solar_position.xls. The calculations are illustrated in the following figure:

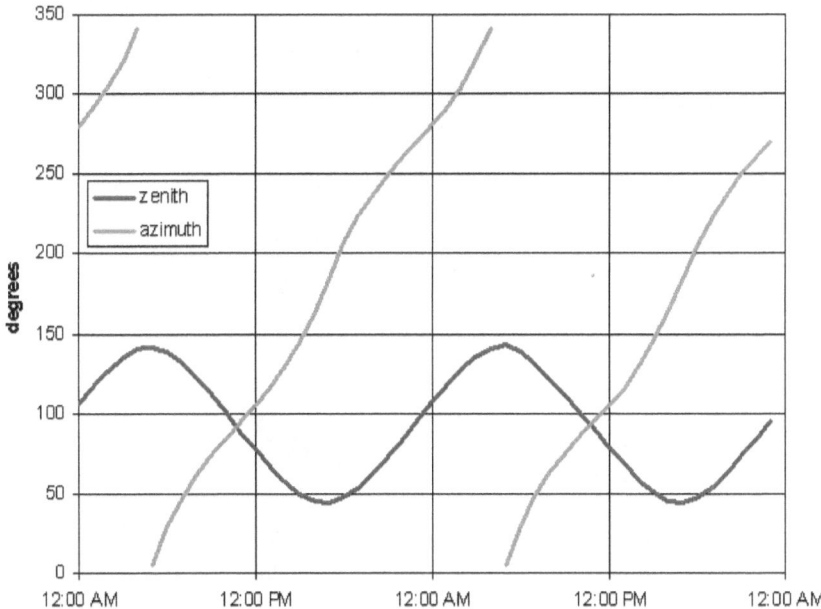

A greatly simplified version of this algorithm can be found in solar.c in this same folder.

```
double RE=6371.01;    /* radius of Earth */
double AU=149597890.; /* astronomical unit */
typedef struct{double z,a;}SP;
SP SolarPosition(int yr,int mo,int da,int hr,int mn,int
   sec,double lon,double lat)
   {
   int i,j;
   double a,b,c,d,e,g,h,l,o,r,s,t,w;
   static SP sp;
   h=hr+(mn+sec/60.)/60.;
   i=(mo-14)/12;
```

99

```c
j=(1461*(yr+4800+i))/4+(367*(mo-2-12*i))/12-
  (3*((yr+4900+i)/100))/4+da-32075;
s=j-0.5+h/24.;
d=s-2451545.;
w=2.1429-0.0010394594*d;
l=4.895063+0.017202791698*d;
b=6.24006+0.0172019699*d;
e=l+0.03341607*sin(b)+0.00034894*sin(2.*b)-0.0001134-
  0.0000203*sin(w);
o=0.4090928-0.000000006214*d+0.0000396*cos(w);
r=atan2(cos(o)*sin(e),cos(e));
if(r<0.)
  r=r+2.*M_PI;
c=asin(sin(o)*sin(e));
g=6.6974243242+0.0657098283*d+h;
t=(g*15.+lon)*(M_PI/180.);
a=t-r;
sp.z=(acos(cos(lat*(M_PI/180.))*cos(a)*cos(c)
  +sin(c)*sin(lat*(M_PI/180.))));
sp.a=atan2(-sin(a),tan(c)*cos(lat*(M_PI/180.))-
  sin(lat*(M_PI/180.))*cos(a));
if(sp.a<0.)
  sp.a=sp.a+2.*M_PI;
sp.a/=M_PI/180.;
sp.z=(sp.z+(RE/AU)*sin(sp.z))/(M_PI/180.);
return(sp);
}
```

also by D. James Benton

3D Articulation: Using OpenGL, ISBN-9798596362480, Amazon, 2021 (book 3 in the 3D series).

3D Models in Motion Using OpenGL, ISBN-9798652987701, Amazon, 2020 (book 2 in the 3D series.

3D Rendering in Windows: How to display three-dimensional objects in Windows with and without OpenGL, ISBN-9781520339610, Amazon, 2016 (book 1 in the 3D series).

A Synergy of Short Stories: The whole may be greater than the sum of the parts, ISBN-9781520340319, Amazon, 2016.

Azeotropes: Behavior and Application, ISBN-9798609748997, Amazon, 2020.

bat-Elohim: Book 3 in the Little Star Trilogy, ISBN-9781686148682, Amazon, 2019.

Boilers: Performance and Testing, ISBN: 9798789062517, Amazon 2021.

Combined 3D Rendering Series: 3D Rendering in Windows®, 3D Models in Motion, and 3D Articulation, ISBN-9798484417032, Amazon, 2021.

Complex Variables: Practical Applications, ISBN-9781794250437, Amazon, 2019.

Compression & Encryption: Algorithms & Software, ISBN-9781081008826, Amazon, 2019.

Computational Fluid Dynamics: an Overview of Methods, ISBN-9781672393775, Amazon, 2019.

Contaminant Transport: A Numerical Approach, ISBN-9798461733216, Amazon, 2021.

CPUnleashed! Tapping Processor Speed, ISBN-9798421420361, Amazon, 2022.

Curve-Fitting: The Science and Art of Approximation, ISBN-9781520339542, Amazon, 2016.

Death by Tie: It was the best of ties. It was the worst of ties. It's what got him killed., ISBN-9798398745931, Amazon, 2023.

Differential Equations: Numerical Methods for Solving, ISBN-9781983004162, Amazon, 2018.

Equations of State: A Graphical Comparison, ISBN-9798843139520, Amazon, 2022.

Evaporative Cooling: The Science of Beating the Heat, ISBN-9781520913346, Amazon, 2017.

Forecasting: Extrapolation and Projection, ISBN-9798394019494, Amazon 2023.

Heat Engines: Thermodynamics, Cycles, & Performance Curves, ISBN-9798486886836, Amazon, 2021.

Heat Exchangers: Performance Prediction & Evaluation, ISBN-9781973589327, Amazon, 2017.

Heat Recovery Steam Generators: Thermal Design and Testing, ISBN-9781691029365, Amazon, 2019.

Heat Transfer: Heat Exchangers, Heat Recovery Steam Generators, & Cooling Towers, ISBN-9798487417831, Amazon, 2021.

Heat Transfer Examples: Practical Problems Solved, ISBN-9798390610763, Amazon, 2023.

The Kick-Start Murders: Visualize revenge, ISBN-9798759083375, Amazon, 2021.

Jamie2: Innocence is easily lost and cannot be restored, ISBN-9781520339375, Amazon, 2016-18.

Kyle Cooper Mysteries: Kick Start, Monte Carlo, and Waterfront Murders, ISBN-9798829365943, Amazon, 2022.

The Last Seraph: Sequel to Little Star, ISBN-9781726802253, Amazon, 2018.

Little Star: God doesn't do things the way we expect Him to. He's better than that! ISBN-9781520338903, Amazon, 2015-17.

Living Math: Seeing mathematics in every day life (and appreciating it more too), ISBN-9781520336992, Amazon, 2016.

Lost Cause: If only history could be changed..., ISBN-9781521173770, Amazon, 2017.

Mass Transfer: Diffusion & Convection, ISBN-9798702403106, Amazon, 2021.

Mill Town Destiny: The Hand of Providence brought them together to rescue the mill, the town, and each other, ISBN-9781520864679, Amazon, 2017.

Monte Carlo Murders: Who Killed Who and Why, ISBN-9798829341848, Amazon, 2022.

Monte Carlo Simulation: The Art of Random Process Characterization, ISBN-9781980577874, Amazon, 2018.

Nonlinear Equations: Numerical Methods for Solving, ISBN-9781717767318, Amazon, 2018.

Numerical Calculus: Differentiation and Integration, ISBN-9781980680901, Amazon, 2018.

Numerical Methods: Nonlinear Equations, Numerical Calculus, & Differential Equations, ISBN-9798486246845, Amazon, 2021.

Orthogonal Functions: The Many Uses of, ISBN-9781719876162, Amazon, 2018.

Overwhelming Evidence: A Pilgrimage, ISBN-9798515642211, Amazon, 2021.

Particle Tracking: Computational Strategies and Diverse Examples, ISBN-9781692512651, Amazon, 2019.

Plumes: Delineation & Transport, ISBN-9781702292771, Amazon, 2019.

Power Plant Performance Curves: for Testing and Dispatch, ISBN-9798640192698, Amazon, 2020.

Practical Linear Algebra: Principles & Software, ISBN-9798860910584, Amazon, 2023.

Props, Fans, & Pumps: Design & Performance, ISBN-9798645391195, Amazon, 2020.

Remediation: Contaminant Transport, Particle Tracking, & Plumes, ISBN-9798485651190, Amazon, 2021.

ROFL: Rolling on the Floor Laughing, ISBN-9781973300007, Amazon, 2017.

Seminole Rain: You don't choose destiny. It chooses you, ISBN-9798668502196, Amazon, 2020.

Septillionth: 1 in 10^{24}, ISBN-9798410762472, Amazon, 2022.

Software Development: Targeted Applications, ISBN-9798850653989, Amazon, 2023.

Software Recipes: Proven Tools, ISBN-9798815229556, Amazon, 2022.

Steam 2020: to 150 GPa and 6000 K, ISBN-9798634643830, Amazon, 2020.

Thermochemical Reactions: Numerical Solutions, ISBN-9781073417872, Amazon, 2019.

Thermodynamic and Transport Properties of Fluids, ISBN-9781092120845, Amazon, 2019.

Thermodynamic Cycles: Effective Modeling Strategies for Software Development, ISBN-9781070934372, Amazon, 2019.

Thermodynamics - Theory & Practice: The science of energy and power, ISBN-9781520339795, Amazon, 2016.

Version-Independent Programming: Code Development Guidelines for the Windows® Operating System, ISBN-9781520339146, Amazon, 2016.

The Waterfront Murders: As you sow, so shall you reap, ISBN-9798611314500, Amazon, 2020.

Weather Data: Where To Get It and How To Process It, ISBN-9798868037894, Amazon, 2023.

www.ingramcontent.com/pod-product-compliance
Lightning Source LLC
Chambersburg PA
CBHW030948240526
45463CB00016B/2157